AF553895

Fisheries, Aquaculture and Biotechnology

Shagufta

APH PUBLISHING CORPORATION
4435-36/7, Ansari Road, Darya Ganj
New Delhi-110 002

Published by
S.B. Nangia
A P H Publishing Corporation
4435-36/7, Ansari Road, Daryaganj
New Delhi 110002
Ph.: 23274050
E-mail : aphbooks@gmail.com

2023

Printed at
Balaji Offset
Navin Shahdara, Delhi 110032

Preface

Biotechnology provides powerful tools for the sustainable development of aquaculture, fisheries, as well as the food industry. Increased public demand for seafood and decreasing natural marine habitats have encouraged scientists to study ways that biotechnology can increase the production of marine food products, and making aquaculture as a growing field of animal research.

Biotechnology allows scientists to identify and combine traits in fish and shellfish to increase productivity and improve quality. Scientists are investigating genes that will increase production of natural fish growth factors as well as the natural defence compounds marine organisms use to fight microbial infections. Modern biotechnology is already making important contributions and poses significant challenges to aquaculture and fisheries development. It perceives that modern biotechnologies should be used as adjuncts to and not as substitutes for conventional technologies in solving problems, and that their application should be need-driven rather than technology-driven.

The use of modern biotechnology to enhance production of aquatic species holds great potential not only to meet demand but also to improve aquaculture. Genetic modification and biotechnology also holds tremendous potential to improve the quality and quantity of fish reared in aquaculture. There is a growing demand for aquaculture; biotechnology can help to meet this demand. As with all biotech-enhanced foods, aquaculture will be strictly regulated before approved for market. Biotech aquaculture also offers environmental benefits. When appropriately integrated with other technologies for the production of food, agricultural products and services, biotechnology can be of significant assistance in meeting the needs of an expanding and increasingly urbanized population in the next millennium. Benefits offered by the new technologies cannot be fulfilled without a continued commitment to basic research. Biotechnological programmes must be fully integrated into a research background and cannot be taken out of context if they are to succeed.

The present volume explores the science and application of biotechnologies to fisheries and aquaculture. It provides various approaches in biotechnology and their potential impact on global fisheries and aquaculture. This book will be most useful for advanced students and researchers in fisheries and aquaculture.

Editor

Contents

1 Introduction

Aquaculture is an ancient form of farming dating back 2000 years or more in China and the Roman Empire. However, only in the last few decades has aquaculture grown into a global practice resulting in tremendous worldwide production. Aquaculture production has enlarged dramatically since the early 1980s, and will become increasingly important as demand for fish products increases, world harvest by capture fisheries reaches a plateau or declines and human population numbers expand. The biomass of fish that can be produced per surface area is much greater than that for terrestrial animals, indicating that aquaculture could be the key for providing global food security.

Humans were hunter-gatherers prior to being farmers, and fishermen before they were aquaculturists. Although aquaculture is growing in importance and must expand to meet future demand for fish products, commercial harvest of natural populations has always been and is still of higher economic value than aquaculture. Even as aquaculture closes the gap or surpasses the value of commercial fisheries, the genetic management and conservation of natural fish stocks and gene pools will be of great importance. Genetic variation is one key variable in the survival of various species. Also, natural populations are perhaps the best gene banks, a critical resource for genetic variation for current and future application in genetic improvement for farmed species and specialized sport-fish applications.

Recreational fishing is also of great importance in many countries. When the revenue from the fish, the licences, fishing equipment, boats, travel, food and lodging is considered, recreational fishing is probably more than tenfold more important economically than aquaculture in the USA. Biotechnology is permanently linked not only to aquaculture, but also to commercial and recreational fisheries because of its potential positive and negative impacts on these resources.

Currently, the quantity of animal protein harvested from global aquatic sources via the capture of natural fish populations is at maximum sustainable yield. Many major fish stocks are showing precipitous declines in productivity due to overfishing and further increases are not anticipated under the current global conditions and environment.

History of Biotechnology in Aquaculture and Fisheries

Wild fish stocks have been heavily fished or overfished, which has resulted in a noticeable levelling of fish landings at around 60 million t, with harvest from oceans unlikely to expand. Almost two-thirds of marine stocks in the Pacific and Atlantic Oceans are being fully exploited or have already been overfished. With increased demand for aquacultured foods has come a need for more efficient production systems. Major improve-ments have been achieved through enhanced husbandry procedures, improved nutrition, enhanced disease diagnosis and therapies and the application of genetics to production traits.

Although several aquaculture species have been greatly improved through the application of genetics, much greater improvements can be accomplished. Genetics can greatly contribute to production efficiency, enhancing production and increasing sustainability. Resource utilization can be greatly improved and impediments to sustainability, such as slow growth of fish, inefficient feed conversion, heavy mortality from disease and the associated use of chemicals, loss of fish from low oxygen levels, inefficient harvest, poor reproduction, inefficient use of land space and processing loss, can all be diminished by utilizing genetically improved fish.

Genetic enhancement of farmed fish has advanced to the point that it is now having an impact on aquaculture worldwide, but potential maximum improvement in overall performance is not close to being achieved. As space for aquaculture becomes more limiting, the necessity for more efficient production or increased production within the same amount of space will further increase the importance of genetic improvement of aquaculture species.

When the Chinese, Europeans and others observed mutations and phenotypic variation for colour, body conformation and finnage, and then selected for these phenotypes as well as for body size, selective breeding, the predecessor of molecular genetics and biotechnology, of fish and shellfish was born. Additionally, fish culturists and scientists who compared and evaluated closely related species for their suitability for aquaculture application over the past two millennia were also unknowingly conducting some of the first fish genetics research. Closely related species are reproductively isolated and have species status because of their genetic distance from one another; therefore, the comparison of different species is a genetic comparison.

The foundation for electrophoretic analysis was laid in 1816 when R. Pornet reported the effect of electric fields on charged particles, including proteins. Brown recognized the nucleus as a regular, constant cellular element within cells in 1831, and in the 1840s Carl Nageli had observed that the nucleus divided first in dividing cells but did not understand the significance of this observation. Although Charles Darwin was not the first to develop the theory of evolution, based on selection of the fittest, this key concept was made believable in the 1860s by Charles Darwin and Alfred R. Wallace. Darwin's grandfather, Erasmus Darwin, was a proponent of evolution, and by the late 18th century Buffon and Lamarck had theorized that acquired characteristics were heritable. Buffon and Lamarck believed that the external environment brought about change, but Geoffroy Saint-Hilaire felt change was embryonic or germinal. Although their knowledge is not recorded in writing, obviously early fish breeders understood these basic concepts.

The relationship between genes and proteins was first suggested by Archibold E. Garrod in 1908. Also that year, G. Hardy and W. Weinberg independently developed some of the basic laws governing population genetics. Thomas Hunt Morgan and Calvin Bridges provided experimental proof of the chromosome theory in 1910. Thomas Hunt Morgan was one of the first to demonstrate the concept of linkage in *Drosophila.* Crossing over was first described by FA. Janssens about 1909 and then verified in 1931 by Barbara McClintock and Harriet Creighton. Early in the 20th century, Thomas Hunt Morgan and his student Alfred Sturtevant described single and multifactorial inheritance, chromosome mapping, gene linkage and recombination, sex linkage, mutagenesis and chromosome aberrations. Sturtevant described linear linkage in 1913. During the 1920s and 1930s much progress was made in the field of population genetics. These efforts were primarily led by R.A. Fisher and Sewell Wright. Much of the research was related to selection, inbreeding and relatedness of individuals and populations, and also application and relevance to quantitative genetics and selective breeding. Also, Embody and Hayford conducted some of the first fish genetics research, a strain comparison of rainbow trout, *Oncorhynchus mykiss,* during this time period.

George Beadle and Edward Tatum advanced the hypothesis of one gene-one enzyme in 1941 and Avery, McCarty and MacLeod indicated that DNA was the physical material for heredity, bringing together Mendelian genetics, biochemistry and cytogenetics for the beginning of molecular genetics. By 1951, Barbara McClintock had identified movable control elements, but the understanding and appreciation of this concept would wait for many years. A major milestone was accomplished in 1953 when James Watson, Francis Crick and Maurice Wilkins discovered the molecular model for the chemical structure of DNA, the double-helical nature of DNA. Between 1961 and 1964, Marshall Nirenberg, Henry Matthaei, Severo Ochoa, H.G. Khorana and others deciphered the genetic code, and then Charles Yanofsky and Alan Garen followed with genetic evidence for the code. Isozyme analysis was also developed in the 1950s. Oliver Smithies developed starch gel electrophoresis, and Clement Markert and R.L. Hunter developed histochemical staining for the visualization of enzymes and isozymes. By the 1960s most of the key components for modern biotechnology were in place.

A Hungarian engineer defined biotechnology as all lines of work by which products are produced from raw materials with the aid of living things, in reference to an integrated process of using sugarbeets to produce pigs. However, the term was associated with industrial fermentation or ergonomics until 1961, when Carl Goran Heden's scientific journal led to biotechnology being defined as the industrial production of goods and services by processes using biological organisms, systems and processes. In the last two decades, biotechnology has often been associated with recombinant DNA technology, but in actuality recombinant DNA technology is a subdiscipline of biotechnology. Fish genetics programmes first emerged in the 1900s after the basic principles of genetics and quantitative genetics had been established. However, there was not a substantial effort in fish genetics research and the application of genetic enhancement programmes until the 1960s because of the infancy and small scale of aquaculture, a lack of knowledge of fish genetics and a lack of appreciation of genetic principles by natural-resource managers regarding genetic enhancement, population genetics and conservation genetics. Slightly earlier, Ellis Prather conducted selection of largemouth bass, *Micropterus salmoides,* for improved feed conversion efficiency in the 1940s and Donaldson selected rainbow trout for increased growth in the 1950s, but neither utilized genetic controls, making any genetic progress unverifiable. Since the 1960s, fish genetics research and application of genetically improved fish and genetics principles have been gaining momentum with each passing decade.

The next major technological breakthrough was the first isolation of restriction endonucleases by Werner Arber, Hamilton O. Smith and Daniel Nathans around 1970, which was the key discovery allowing the development of gene cloning, genetic engineering and various restriction-fragment technologies. The discovery of reverse transcriptase by Howard Temin and David Baltimore was, of course, also key for the development of modern recombinant DNA technology. Then in 1973, Stanley Cohen and Herbert Boyer devised recombinant DNA technology. This type of research was further enhanced in 1975 with the development of procedures to rapidly obtain DNA sequences and to visualize specific DNA fragments. The 1980s saw more quantum leaps in molecular genetic biotechnology. Around 1980, Palmiter, Brinster and Wagner produced the first transgenic animals, mice, and Palmiter and Brinster demonstrated that the transgenesis could lead to greatly accelerated growth in the mice. Palmiter, Brinster and Wagner demonstrated the dramatic pheno-typic alterations that could be realized through gene transfer. This provided the motivation and impetus for the development of technology for the generation of the first transgenic fish. In a year-and-a-half span from 1985 to 1987, Zhou first transferred genes into goldfish in China, followed by Ozato in Japan with medaka, Daniel Chourrout in France with rainbow trout and Rex Dunham in the USA with channel catfish.

The new biotechnologies, such as sex reversal and breeding and polyploidy, began to have a major impact on aquaculture production in the late 1980s and early 1990s by not only improving growth rates but allowing major improvement of flesh quality in species that exhibit sexual dimorphic and sexual maturation effects. Chourrout induced the first viable tetraploid fish, rainbow trout, Standish Allen developed triploid technology for shellfish during the late 1980s and Gary Thorgaard developed clonal lines of rainbow trout via androgenesis. The pioneering research on sex reversal and breeding technology by Shelton and Guerrero led to worldwide production of monosex Nile tilapia in the 1980s and 1990s, and Graham Mair took this technology one step further in the 1990s, leading to the development of YY populations of Nile tilapia and the production of genetically male tilapia (GMT) populations in many countries. Traditional breeding has already been utilized in concert with these new biotechnologies.

Environmental concerns about the application of biotechnology and genetic engineering emerged in the 1980s. Application of gene-transfer technology will not happen until genetic engineering is proved to be a safe technology. In the mid-1990s, Dunham conducted the first environmental-risk research with transgenic fish, channel catfish, demonstrating that in natural conditions the transgenics were slightly less fit than non-transgenic cohorts. Also in the mid-1990s, Du, Choy Hew, Garth Fletcher and Robert Devlin produced the first transgenic fish, salmon, exhibiting hyperlevels of growth -two- to sixfold and, in the case of Devlin's research, 10-30-fold increases in growth rate. Shortly thereafter, Norman Maclean produced transgenic tilapia with a two- to fourfold increase in growth rate.

In the last few years, technological advances in DNA marker technologies and DNA microarray and gene chip technologies have further accelerated the pace of aquaculture and aquaculture genomics. Genomic research has produced vast amounts of information towards an understanding of the genomic structures, organization, evolution and genes involved in the determination of important economic traits of aquatic organisms. Positional cloning of genes from aquatic species is no longer a dream. Since the early 1980s, research in aquaculture and fisheries genetic biotechnology has steadily grown, and now research in this area is extremely active. Currently, efforts are well established in the areas of traditional selective breeding, biotechnology and molecular genetics of aquatic organisms. Cultured fish are being improved for a multitude of traits, including growth rate, feed conversion efficiency, disease resistance, tolerance of low water quality, cold tolerance, body shape, dress-out percentage, carcass quality, fish quality, fertility and reproduction and harvestability. For many years there has been a cry in the wilderness that aquaculture is impeded by the lack of genetically improved fish and the utilization of essentially wild fish. This is still true for some species and for new aquaculture species; however, for a few well-established aquatic species, large genetic gain has been realized, and there is evidence of up to tenfold improvement of some traits compared with that of poor-performing, unimproved wild strains by combining various combinations of traditional selective breeding and biotechnology. The development and utilization of genetically improved fish are widespread across the world in the 21st century. A variety of genetic techniques are being implemented commercially, including domestication, selection, intraspecific crossbreeding, interspecific hybridization, sex reversal and breeding and polyploidy, to improve aquacultured fish and shellfish. Genetically improved fish and shellfish from several different phylogenetic families are utilized. Genetic principles and biotechnology are also being utilized by fisheries managers and by researchers to enhance natural fisheries, to protect native populations and to genetically conserve natural resources. Genetically modified aquatic organisms are already having an impact on global food security in both developed and developing countries. However, in general, much more progress can and needs to be made.

Applications of Biotechnology in Fisheries and Aquaculture

Currently, there is a beneficial interaction between biotechnology and aquaculture. The first is conceived as a set of techniques used or applied to living organisms aimed at modifying or manufacturing a product for use or consumption. Aquaculture is defined as the production of aquatic organisms under controlled systems with technological applications that allow managing population densities higher than the natural ones, optimizing culture management; in other words, an intensive aquaculture.

Biotechnology started by contributing with recombinant DNA which enables a closer interaction and greater knowledge of the genetics of the species and aquaculture has quantitatively increased over the last five decades due to contributions made by different fields of knowledge. Biotechnology is the most relevant one in areas such as reproduction, nutrition, pathology, and genetic improvement of cultured species. Biotechnology can provide aquaculture with the means to increase significantly its intensity and capacity over the next few years.

Biotechnology to Improve Reproduction

Reproductive Maturity Control

Sexual maturity is a lengthy process that in different organisms, it often represents profound physiological and morphological changes, which in turn may alter the quality of the products obtained from them. This affect the timing of the final stages of cultivation, particularly it determines harvest time. In animals, the main objective of controlling reproduction through reproductive biotechnologies is the temporary control of maturation, fertility, genetic sex and the sex ratio of the progenies.

Fish have a reproductive cycle that it is understood as the period of the year when the natural spawning of the species takes place. This period is normally short, with a duration and timing according to the species. However, spawning occurs after gonads have been developed which is a lengthy process. The reproductive cycle of salmonids and other fish, includes puberty, gonadal recrudescence, and sexual maturity. These events are modulated by environmental factors and controlled by their endocrine system, allowing fish to produce gametes and have descendants. The salmonids introduced in the southern hemisphere start the recrudescence period after being stimulated by increasingly longer photoperiods after winter solstice which stimulates neuroendocrine activity, and finally spawn at the end of autumn or at the beginning of the next winter. These photoperiodic and water temperature variations are photic and thermal information that fish capture through their eyes. The pineal gland, with photoreceptive capacity, also perceives light stimuli that are sent to the hypothalamus which transforms them into chemical information and pulses that secrete neurotransmitters and synthesize hormones.

The neuroendocrine system comprises a group of hormones from the hypothalamus, hypophysis, and gonads. The hypothalamus plays a regulating role by stimulating and inhibiting the hypophysis through Gn-RH hormone and dopamine, respectively. The hypophysis releases gonadotropin GTH that is the most important hormone for oocyte maturation, and it is present as GTH-I and GTH-II. The GTH-I in the ovary acts on the thecal and follicular granulose cells to synthesize steroid 17 β estradiol, which in turn acts on the liver to start and sustain the synthesis of vitellogenin in the oocyte. GTH-II captures serum vitellogenin to include it in the oocyte. 17 β estradiol releases gonadotropin, intervenes in the gonadal maturation processes, and stimulates the development of secondary sexual characteristics of the fish. In the final oocyte maturation phase GTH-I level rises, stimulating thecal cells to produce 17 α 20 β dihydroxyprogesterone which takes part in the haploidization, prior to ovulation.In males, GTH-I acts on the Leyding cells which trigger production of 11 keto-testosterone; the hormone that induces testicular maturation via spermatogenesis. This hormone is also related to male secondary sexual characteristics and its maximum production coincides with spermiation. The highest serum levels of GTH-I, are also reached during spermiation, stimulating the production of 17 α 20 β dihydroxyprogesterone to control the transport of sodium and potassium being the latter required to keep sperm cells immobilized.

The endocrinological knowledge has been crucial for the development of useful biotechnologies that allow to manage the reproductive maturation, to advance or postpone this process, accelerating or delaying the ovulation and spermiation, and therefore to manage the availability of oocytes and sperms. Fish endocrine functioning has been described by Zanuy and Carrillo. For the main farmed salmonids, rainbow trout (*Oncorhynchus mykiss*), coho salmon (*Oncorhynchus kisutch*), and brown trout (*Salmo trutta*), have published a series of papers which are the basis to apply reproductive biotechnologies.

In commercial fish farming, natural maturation forces dependence on gamete availability at a given time, and this also condition harvesting time. In the event of forcing maturation, it is possible to obtain advantage by shortening the time to market, instead, by delaying maturation it is possible to have more time for fish to grow and eventually culture bigger smolts (final phase of smolt farming when fish are ready to be released into the sea water). Through the combination of alternatives to bring forward or delay maturation it is also possible to manage advanced, normal, and delayed spawning stock in the hatchery, cover most of the year with the production of eggs and harvests, and have a greater productive efficiency by making a better use of the infrastructure and staff.

Two biotechnologies, physiological in nature, are available as part of those devoted to commercial egg production; and are used on fish selected to restock breeders. These technologies affect the functioning of the hypothalamus-hypophysis-gonad axis and as such, influence the physiology of the reproducer, modifying maturation time, without neither passing this on to their descendants, nor repeating the phenomenon on the same breeders in next maturation period. First, the environmentally induced maturity, based on the external control of the photoperiod, to induce maturity that should be applied before the beginning of such process. Secondly, the hormonal induced maturity which is based on the administration of natural or synthetic hormones such as GnRH slightly before final fish maturation.

Photoperiod Management

Numerous studies report the use of artificial photoperiods to modify, among other, fish maturation time. One of the oldest reports was published by Hoover who applied it to speed up the sexual cycle of brook trout (*Salvelinus fontinalis*). Most of the studies have been done on rainbow trout (*Oncorhynchus mykiss*) and other salmonids, however, there are also information on several flatfish such as the Atlantic halibut, *Hippoglossus hipoglossus,* turbot, *Scphthalmus maximus* common sole, *Solea solea* and *Paralichthys dentatus*. Other fish that have been studied include the *European sea bass, Dicentrarchus labrax, red drum, Sciaenops ocellatus, Atlantic cod, Gadus morhua, goldfish, Carassius auratus, gilthead seabream, Sparus aurata,* barbo, *Barbus barbus,* and *striped trumpeter, Latris lineata.*

Accelerated photoperiod has been commonly used to advance fish maturation but, it is also possible to delay it with extended photoperiod regimes. For instance, a significant advance of spawning in rainbow trout *(Oncorhynchus mykiss)* can be obtained with accelerated photoperiod regimes from 6 to 9 months, and in contrast, a spawning delay can be achieved with a continuous light regime.

There are other possible uses of the photoperiod, for instance, Clarke et al. demonstrated that coho and chinook salmons continuously exposed to short day light regimes for two months at onset of feeding can uniformly produce smolts SO (smolts prior to the first year of life). Beacham described that short photoperiods, applied for two months, improved smoltification of the Atlantic salmon in autumn and also produced better survival and growth in winter. Sigholt *et al.* described that exposure of one-year old Atlantic salmon fingerlings during 5-7 weeks of short day light followed by continuous light, triggered changes related to smoltification.According to Clarke the growth of coho salmon, but not necessarily of chinook salmon, was considerably higher when they were exposed to changing photoperiod vs. a continuous 12 h day light period. Forsberg reported that the post-smolt growth of the Atlantic salmon was strongly influenced by photoperiod in ground farms. The most recent and most complete updated information on the effect of photoperiod on maturation regulation was published by Bromage *et al.* who reviewed, on fish, the effect of the interaction between the photoperiod with other environmental factors such as temperature, salinity, nutritional status and melatonin.

Hormonally-induced Ovulation

In a recent review, Zohar and Mylonas summarized the progress made in the use of hormonal methodologies, that started with the work done by Houssay demonstrating the effects of hypophyseal extracts on the sexual maturation of fish and reptiles. Based on these results, the hypophyzation, a technique that has been applied to fish and it consists of injecting fish with hypophyseal extracts containing, among other products, hormones for sexual maturation was developed. Since the seventies, there are hypophyseal preparations, chromatographically purified from carps and salmons available for these and other phylogenetically related species. Also during the seventies, began the use of the human chorionic gonadotropin (hCG) to control fish maturity, available without restrictions as clinical preparations. During the same time, the use of gonadotropin release hormones (GnRH) started, first in mammals and later in fish. Finally, the latter were chemically synthesized, and are known as analogs or GnRHa. They are cheaper and more efficient in inducing maturation, can be injected or implanted as a pellet.

Zohar and Mylonas also include species of *Carassius, Clarias, Cyprinus, Oreocrhromis, Perca, Solea* y *Sparus*. Therefore, several species showed positive responses to exogenous hormone administration, being possible to anticipate the ovulation and spermiation. However, this response to hormone induction may have some negative effects on the quality of the gametes obtained so, it is necessary to evaluate this situation before to use it.

Fertility Control

The changes associated to the reproductive maturation already analyzed also justify the importance of controlling fertility in fish cultures. In general, it is postulated that physiological and morphological changes related to maturation imply devoting a great deal of metabolic energy to these activities. This may stops growth, or at least may reduce growth rate, from the beginning of maturation. Then, it is possible that by blocking maturation and getting sterile individuals without gonadal development, will sustain a high growth rate, a larger size at harvesting without change in color or lost other meat quality due to the movement of lipids from muscles to gonads and because of the subsequent hydration of muscle tissue. Among the commonest alternatives to induce sterility, the use of polyploidyzation has been tested in many species.

Experimental Polyploidy

Polyploidy is a quantitative change of the genome affecting all the chromosomes in the cell, increasing the number of chromosomic sets. Experimental polyploidy induction in fish is carried out by hampering meiosis II in the female once the oocyte is fertilized. This prevents the chromosomes reduction to a half (n). So, the embryo will be triploid (3n), formed by two sets of chromosomes from the mother, and one from the father. This is a low cost methodology, highly efficient, with low mortality rate, and the only disadvantage is the poor response of males, but it is very advantageous when only female populations are managed. Experimental methods used are:

1. physical methods, e.g. thermal shocks of heat or cold, pressure or thermoelectric shocks,
2. chemical methods, e.g. applications of cytochalasins B, colchicine or 6-dimethylaminopurine, and
3. genetic methods, e.g. mating diploid females and tetraploid males.

In fish, the first successful triploidy induction in salmonids was reported by Svardson and numerous reports describing polyploidy induction in fish of economic importance have been published since 1970. The production of polyploid individuals has various possible applications. As for triploids, that are totally or partially sterile, the application may allow to increase meat production efficiency by reducing energy consumption during the process of sexual maturity, boosting growth, avoiding deterioration of meat quality and preventing changes in the color of the skin. By avoiding maturation, harvests can also be delayed thus, getting bigger and older fish. It is also possible to avoid precocious males from appearing among fish intended for harvesting. Tetraploids have normal fertility, and are potentially useful in the production of triploids by mating with normal individuals. These are obtained by hampering the first mitotic egg segmentation. By avoiding cell division (cytodieresis), nucleus of the embryo is formed by doubling the chromosomes of the species, giving origin to a tetraploid embryo (4n).

Polyploidy Induction

Polyploidy induction has been applied to numerous fish species. For instance, the bibliographic review of triploidy in these organisms from 1943 to 1988 published by Benfey, revealed work done in *Salmonidae,* Characidae, Cyprinidae, Centrarchidae, *Ciclilidae, Pleuronectidae* families. Recent work including turbot (*Scophthalmus maximus*) analyzed the impact of temperature, and the duration of cold shocks on larvae survival and triploidy rates. At the same time, it validated the use of regions of nucleolar organization analysis (NOR analysis) for checking ploidy levels in this species. The cold thermal shocks in triploid inductions at commercial level analyzing the effect of the duration of cold thermal shock on survival and triploidy percentages, in order to obtain 100% of triploids with the best larvae survival have been used, and comparisons of early survival and growth of triploids v/s diploids. Triploidy induction has also been applied to yellow tail common sole (*Pleuronectes ferrugineus*). For instance, Manning and Crim researched the parameters required to produce triploid larvae by applying post fertilization hydrostatic pressure. In seabream (*Sparus aurata*), Haffray *et al.* demonstrated that triploid fish show the same growth rate as diploids, with no difference in terms of survival, yield, and quality. The same is valid for the red seabream (*Pagrus major*).

Based on the published information, it becomes apparent that there is no significant advantage in terms of growth from triploid adult size fish. Nevertheless, some productive assessments in salmonids conclude that a better juvenile growth is achieved with triploid fish when early maturation is controlled. In some cases, better-feed conversions with cost reduction, and fish with good meat quality for processing and smoking are attained. Arai has reviewed the use of chromosomic manipulation techniques of fish in Japan, including triploidy induction in species such as rainbow trout, amago salmon (*Oncorhynchus rhodurus*), masu salmon (*Oncorhynchus masou*), yamame salmon (*Oncorhynchus masou,* fresh water salmon), and locha (*Misgurnus anguillicaudatus*).

According to Gosling, in bivalve mollusks the application of chromosomic manipulation begun at the beginning of the 80s, and triploidization advantages being very similar to the ones already described for fish. The main advantages are the higher somatic growth due to a reassigning of the metabolic energy and avoiding deterioration of meat quality due to maturation. In bivalves, it is possible to induce triploids by blocking meiosis I or II since gametes are released prior the extrusion of the first polocyte. This is obtained using physical or chemical treatment similar to the ones already described for fish. Also, tetraploids can be produced by blocking the first mitotic division.

Sex Proportion Control

The period of time that organisms can be kept in culture is determined by their sexual maturation. At the beginning of this biological process, profound morphological and physiological changes occur. In fish, changes of skin color and meat quality due to the movement of lipids from muscles to gonads and the subsequent hydration of muscle tissue have been described. In some species, the early maturation of an important part of the population, usually males, is an additional problem. Another important aspect is the difference that may exist between the growth rate of males and females of the same species. This could determine that the size achieved within a certain period, be very different and lead to some preferences from the productive standpoint. If fish of a specific sex grow fast and/or mature late, the tendency would be to favor the management of that specific sex. For example, if female maturity is late, the alternative could be to stay longer in seawater and gain weight, or postpone harvest until market price is high enough. In order to face this type of situations, manipulation techniques have been developed. These can be direct at the level gonadal physiology, or indirect, by modifying sex genetically which allows modifications in the reproductive process aimed at farming monosex fish.

There are two ways of giving hormones: by immersion or ingestion. The latter is the commonest way, where the hormone is included in the normal diet, and fish are fed during a given time. Variables to be considered are the nature of the hormone, diet concentration level, duration of intake, and time of treatment initiation. In the event of producing only male stocks by applying masculinization technique with the use of sexual hormones, there are certain androgynous hormones of natural origin such as testosterone, 11 ketotestosterone, and androstenedione. Among the synthetic ones, 17 α-methyltestosterone is widely used which is, easy to get, chemically stable, but not necessarily the most powerful.

The hormone is added to the diet, dissolved in absolute ethanol that rapidly evaporates. Concentration ratio has been determined by trial and error assays, but the trend is to use the lowest concentration to obtain an appropriate change level. The normal dosage used in rainbow trout is 3 mg·kg^{-1} of feed, which approximately gives 100% of new males from genotypic females. Dosages of 1.0 and 0.5 mg·kg^{-1} of feed also produce a good percentage of new males (80%). For rainbow trout, the best results are obtained if treatment begun at the onset of the first feeding. In coho salmon, the treatment should begin when 50% of the eggs are hatching, and treatment should be given by immersing eggs in a 400 μ·L^{-1} of hormone solution.

As far as treatment duration is concerned, at the concentrations given for rainbow trout, best results are obtained after 60 to 90 days of treatment. From 120 days up, high percentages of sterility are obtained. This also occurred at higher hormone concentrations. In coho salmon, the highest male proportion is produced at a single immersion at 514 degree days in a 400 μ de 17 α -methyltestosterone solution during two hours, producing 78% male.

Based on the results obtained by Yamamoto, fish feminization is also possible with estrogens, producing only female stocks. These treatments are less efficient to revert male to female, and so, it is necessary to provide higher dosage to obtain good reversion rates. Another important difference with masculinization is that oral treatment should be given together with immersion in estrogen solution. The most widely used estrogen is 17 β-estradiol, which is easy to get and chemically stable. It is included in food in an absolute ethanol solution that then evaporates. For salmonids, Purdom *et al.* reported the results of feminization in *O. mykiss, O. kisutch, O. tschawytscha, Salmo trutta,* using different dosages of estradiol per kg feed, or per L of water, and different treatment periods to get optimal reversion results.

Indirect Production of Monosex Progenies

The use of androgens or estrogens produces all male or all female progenies. In Piferrer work, two indirect ways of producing these progenies were analyzed, with the advantage of genetic and physiological sex correspondence among individuals. Likewise, in species where the female is homogametic, all male progenies can be farmed by feminizing normal breeding progenies, part of which will be female converted male (new female). By mating these new females with heterogametic males, a part of homogametic males will be obtained; all male progenies will be produced mating these with homogametic females. In species where the male is homogametic, all male progenies could be obtained by feminizing a normal breeding progenies, part of which will be new females, while mating homogametic males will produced all male progenies.

Likewise, all female progenies can be produced. In species where females are homogametic, all female progenies can be farmed by masculinizing normal breeding progenies, part of which will be male converted female (new male). By mating these new male with homogametic female, all female progenies will occur. Díaz *et al.*, described sex change in rainbow trout in hatcheries to obtain new males, including morphological and chromosomic characterization of new male and all female progeny production.

In species where the female is heterogametic, all female progenies could be produced by masculinizing during the first stage a progeny from regular mating, part of which will result in male converted female (new male). Mating these males with heterogametic females will produce a super female portion that mating with homogametic males will generate all female offspring. *Monosex progeny production via genetic sex control.* It is important to remember that genetic sex determination occurs at fertilization. In some fish species, such as salmonids, genetic sex determination is of XX/XY type, and the differentiation of sexual chromosomes is developed enough to be able to identify these chromosomes when examining the karyotype.

These situations allow the use of mechanisms such as parthenogenesis and artificial induction to control sex genetics in certain breeding progenies. The kinds of parthenogenesis are gynogenesis and androgenesis, both to be experimentally induced. In gynogenesis genetic material of individuals comes only from the mother. Embryo is developed by activating oocytes with genetically inactivated sperm cells. The production of gynogenetics requires combining semen inactivation with diploidyzation of mother chromosomic complement. That is, the egg haploid is first induced, prior sperm cells treatment to eliminate their genetic contribution, and then, eggs are subject to another treatment hampering meiotic division or the first mitotic division to duplicate genetic material.

Natural gynogenesis was first described in *Poecilia formosa* fish, and then in other species of this Family *Poecilidae,* which are gynogenetically reproduced in nature. What really happens is an activation of eggs by heterologous sperm cells, that is, from other species, which, without contributing with their DNA stimulate egg development. Gynogenesis has also been observed in other Families: flatfish (*Pleuronectidae*), sturgeons (*Acipenseridae*), trouts (*Salmonidae*), carps (*Cyprinidae*), catfish (*Siluridae*), and characins (*Characidae*). For many years, there has been information available that renders an account of the experimental induction of gynogenesis in fish: In *Salmo trutta fario,* brown trout, Opperman developed gynogenetic smolts using sperm cells radiated with radium. Later, gynogenesis induction in locha, carp, and other species was reported.

Experimental gynogenesis can not be directly applied to produce all female progenies in hatcheries given the high inbreeding and high mortality rate of progenies produced this way. That is why gynogenetic production is considered related to masculinization, producing new males, and from them, all female progenies. Experimental induction of gynogenesis has been done in rainbow trout, coho salmon, and Atlantic salmon, One of the first ones was the induction of rainbow trout establishing a protocol for that species. It is interesting to add the possibility of producing sterile female monosex fish, to avoid inconveniences of sexual maturity. To produce these fish it is necessary to use new male semen to form the zygote plus the application of thermal shock to produce triploidization. Arai has reviewed the use of chromosomic manipulation techniques to produce fish monosex progenies in Japan, including species such as rainbow trout, amago salmon, masu salmon, yamame salmon, coho salmon, ayu, char, and hirame. All female progeny production is predominant and quite often all female triploids.

Alternatively, androgenesis enables only heredity from the male, and experimentally, may be produced by inactivating egg chromosomes, fertilizing with normal sperm cells, and inhibiting the first egg mitotic division. This methods has been effective in salmonids such as amago salmon, sturgeon, and carp. For genetic determination of sex in tilapia species where males are homogametic, Beardmore *et al.* have produced the androgenetics through successive breeding and feminizations. Androgenetic fish show low survival rate which improves when diploid or tetraploid sperm cells are used whether natural or induced. The main advantage is faster growth and shorter male generational periods in different species.

Biotechnology in Fish Nutrition and Disease Control

Application of classical biotechnology in these two areas is irrelevant in terms of effect on productivity.In the first case, it is worth mentioning the use of growth promoters applied to improve weight and feed conversion in shrimps and salmonids.

Disease problem area major constraint for development of aquaculture. biotechnological tools such as molecular diagnostic methods, use of vaccines and immunostimulants are gaining popularity for improving the disease resistance in fish and shellish species world over for viral diseases, avoidance of the pathogen in very important.in this context there is a need to rapid method for detection of the pathogen. Biotechnological tools such as gene probes and polymerase chain reaction (PCR) are showing great potential in this area. Gene probes and PCR based diagnostic methods have developed for a number of pathogens affecting fish and shrimp. In case of finfish aquaculture, number of vaccine against bacteria and viruses have been developed. Some of these have been conventional vaccines consisting of killed microorgansism but new generation of vaccine consisting of protein subunit vaccine genetically engineered organism and DNA vaccine are currently under development.

In the vertebrate system, immunization against disease is a common strategy. However the immune system of shrimp is rather poorly developed, biotechnological tools are helpful for development of molecule, which can stimulate this immune system of shrimp. Recent studies have shown that the non specific defense system can be stimulated using, microbial product such as lipopolysacharides, peptidoglycans or glucans. Among the immunostimulants known to be effective in fish glucan and levamisole enhance phagocytic activities and specific antibody responses.

Cryopreservation of Gametes or Gene Banking

Cryopreservation is a technique, which involve long-term preservation and storage of biological material at a very low temperature usually at -196 C ,the temperature of liquid nitrogen. It is based on the principle that very low temperature tranquilize or immobilize the physiological and biochemical activities of cell, thereby making it possible to keep them viable for very long period.

The technology of cryopreservation of fish spermatozoa has been adopted for animal husbandary . The first success in preserving fish sperm at low temperature was reported by Blaxter who fertilizes Herring eggs with frozen thawed semen .The spermatozoa of almost all cultivable fish species has now been cryopreserved. Cryopreservation overcomes problems of male maturing before female, allow selective breeding and stock improvement and enables the conservation. One of the emerging requirement for that can be used by breeders for evolving new strains. Most of the plant varieties that has been produced are based on the gene bank collections. Aquatic gene bank however suffers from the fact that at present it is possible to cryopreserve only the male gametes of finfishes and there in no viable technique for finfish eggs and embryos. However , the recent report on the freezing of shrimp embryos. However , the recent report on the freezing of shrimps embryos by subramoniam and newton and Diwan and kandaswami look promising. Therefore, it is essential that gene banking of cultivated and cultivable aquatic species be undertaken expeditiously.

Biotechnological research and development are growing at a very fast rate. The biotechnology has assumed greatest importance in recent years in the development of fisheries, agriculture and human health. The science of biotechnology has endowed us with new tools and tremendous power to create novel genes and genotypes of plants, animals and fish. The application of biotechnology in the fisheries sector is a relatively recent practice. Neverthless,it is a promising area to enhance fish production. The increased application of biotechnological tools can certainly revolutionise our fish farming besides its role in biodiversity conservation.

References

Battacharya S., Dasgupta, S, Datta, M and Basu, D., (2002). Biotechnology input in fish breeding. *Indian journal of biotechnology*, 1:29-38.

Carvalho, G.R. & Pitcher, T.J. (eds) (1995) *Molecular Genetics in Fisheries*. Chapman & Hall, London.

Karunasagar, I. (1999). Diagnosis treatment and prevention of microbial diseases of fish and shellfish. *Curr.Sci.*, 76:387-399.

Lakra, W.S. and Das ,P., (1998). Genetic engineering in aquaculture, *Indian .J.Anim.Sci.*, 68(8);873-879.

2

Genetic Variation in Fish

Genetic variation occurs both within and among populations. Genetic variation is important because it provides the "raw material" for natural selection. Genetic variation is brought about by mutation, a change in a chemical structure of a gene.

Genetic variation among individuals within a population can be identified at a variety of levels. It is possible to identify genetic variation from observations of phenotypic variation in either quantitative traits (traits that vary continuously and are coded for by many genes, e.g., leg length in dogs) or discrete traits (traits that fall into discrete categories and are coded for by one or a few genes. Genetic variation can also be identified by examining variation at the level of enzymes using the process of protein electrophoresis. Polymorphic genes have more than one allele at each locus. Half of the genes that code for enzymes in insects and plants may be polymorphic, whereas polymorphisms are less common in vertebrates.

Ultimately, genetic variation is caused by variation in the order of bases in the nucleotides in genes. New technology now allows scientists to directly sequence DNA which has identified even more genetic variation than was previously detected by protein electrophoresis. Examination of DNA has shown genetic variation in both coding regions and in the non-coding intron region of genes.

Genetic variation will result in phenotypic variation if variation in the order of nucleotides in the DNA sequence results in a difference in the order of amino acids in proteins coded by that DNA sequence, and if the resultant differences in amino acid sequence influence the shape, and thus the function of the enzyme.

Deoxyribose Nucleic Acid (DNA)

The discovery by Watson and Crick of the structure of DNA in 1953 was a landmark in understanding of how genetic information passes from generation to generation. In the half century since then, the fields of molecular biology and genetics have become inextricably linked and developments, particularly over the past 25 years, have opened up the potential of DNA biotechnology. The structure of DNA enables it to carry the information for a cell to reproduce itself. It is a polymeric molecule, that is, made up of a chain of subunits, consisting of chains of nucleotide monomers. Each nucleotide contains a base, along with a sugar (deoxyribose) and a phosphate group (Fig. 1). There are four individual bases, adenine, guanine, thymine and cytosine and they are usually referred to by their first letter abbreviations, A, G, T and C. Two of the bases, A and G, have a double-ring structure and are known as purines. The other two bases, T and C, are pyrimidines with a single carbon-nitrogen ring.

Each nucleotide is a single unit that joins with neighbouring nucleotides in a linear fashion to make up a polynucleotide chain. Particular carbon atoms in the 5-carbon structure of deoxyribose are referred to by numbers, 1' (one prime) to 5'. The link between nucleotides is formed when the 5' of one bonds to the 3' of the next via a phosphodiester bond. It is the sequence of the four bases in a polynucleotide chain which acts as the code for genetic information. The functional beauty of the DNA molecule is a result of complementary base pairing where G can only bond with C, and A can only bond with T, at the middle of the molecule. It means that the two strands are complementary such that the base sequence of one strand predicts and determines the base sequence of the other strand. Because one strand predicts the other it can be used to replicate the sequence.

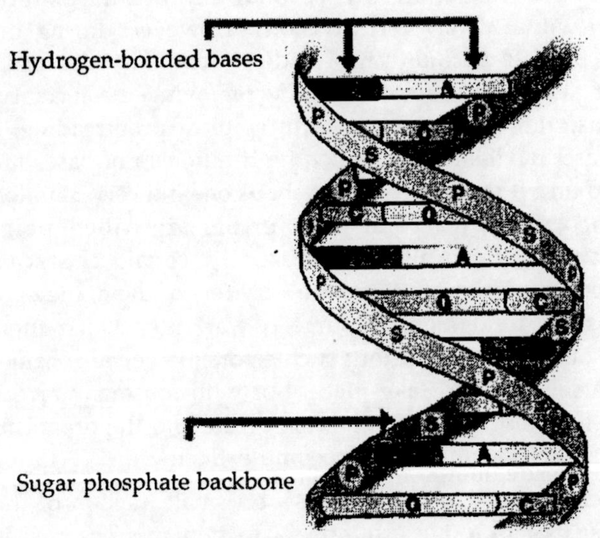

Fig. 1. The structure of DNA

The replication process produces daughter molecules, each of which has one parental strand and one copied strand. This is called semi-conservative replication. Replication of DNA takes place every time a cell divides. The cell's entire DNA is progressively unwound revealing short single-stranded regions which can be copied by DNA polymerase enzymes. Unwinding does not begin at the ends of the molecule, but at points called replication origins, and it then proceeds from these points along the DNA. The new strands of DNA being synthesised during replication are always synthesised in the 5' to 3' direction. This means that as the original strands separate, one new strand can be continuously synthesised against its copy strand (the leading strand) while the other has to be synthesised intermittently in short lengths as enough copy strand (the lagging strand) becomes available (Fig. 2).

Considering the enormous numbers of bases and coded information in the DNA of a cell, replication needs to be extremely accurate. Even a very small incidence of mistakes in copying would result in the loss of important genetic information within a few cell divisions. However, during the replication process various proofreading activities take place and almost all errors are corrected by removing the incorrect base and inserting the correct one. In spite of proofreading, a few errors are inevitable when such high numbers of bases are to be copied and it is estimated that about one in every 3 billion bases is incorrectly inserted. Such errors are called point mutations and they can also be induced by certain chemicals and radioactivity. Although there are very few of them, they are nevertheless the fundamental source of variation which fuels the process of evolution. Without such errors, no genetic change at the DNA level would take place, but with too many errors daughter cells would too often be non-viable and the organism carrying that DNA would soon become extinct.

Functional sequences only represent a small fraction of the total genome, for example around 3% in humans. The rest is made up of what has been called 'junk DNA. Whether all of it is really 'junk' is not known, but it is possible that much of it will have some, as yet undiscovered, function in the organism.

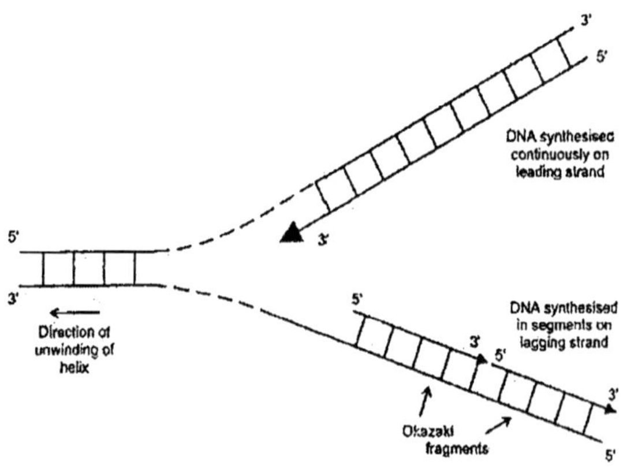

Fig. 2. Replication of DNA.

Some of this junk DNA consists of pseudogenes, genes that for some reason or another have become nonfunctional. Yet other parts of non-coding DNA consist of dispersed or clustered repeated sequences of varying length, from one base pair (bp) to thousands of bases (kilobases, kb) in length. The dispersed repeated sequences occur as copies spread across the genome and can be categorised as long or short interspersed nuclear elements (LINE or SINE), long terminal repeats (LTR) and DNA transposons.

A gene is a unit of information which is held as a code in a discreet segment of DNA. This code specifies the amino acid sequence of a protein. Scientists were surprised to discover quite early on that the sequence information for a single gene was not continuous along the DNA, but was interspersed with pieces of non-coding sequence. The coding parts of a gene sequence are exons, and the non-coding parts are introns (2.3). Before a gene can be expressed, the DNA that encodes it has to be transcribed into RNA.

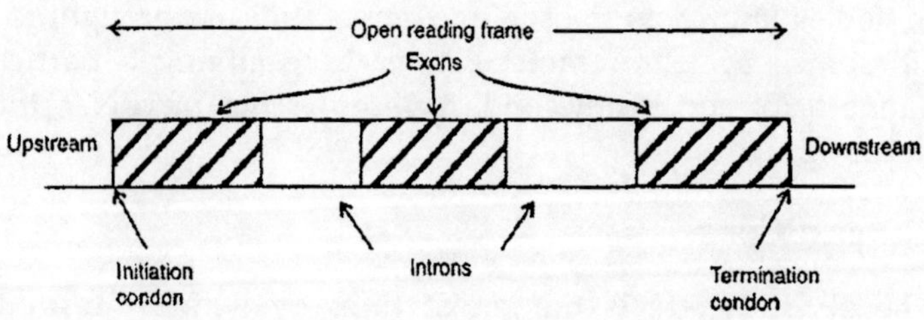

Fig. 3. Generalised structure of a gene.

Ribose Nucleic Acid (RNA)

The structure of ribose nucleic acid (RNA) is similar to that of DNA (deoxyribose nucleic acid) except that (a) the sugar is ribose instead of deoxyribose, (b) in the place of thymine, a similarly structured base called uracil (U) is present and (c) the molecule consists of only a single polynucleotide strand. RNA molecules are produced by the process of transcription of the linear sequence of bases in DNA and are then used in the translation of that sequence into a chain of amino acids that go to make up a protein. The type of RNA transcribed from the sequence is called messenger RNA (mRNA) and translation of this sequence into a string of amino acids is undertaken by ribosomal RNA (rRNA) and transfer RNA (tRNA) molecules.

During transcription of the DNA, an RNA copy is made of one of the strands of DNA. The two strands of DNA are called the template strand and the non-template strand. Other names for the non-template strand are the sense (+) strand or the coding strand. The RNA is synthesised by RNA polymerase enzymes using the template strand and is therefore a copy of the non-template (sense or coding) strand of DNA. Because this RNA is a direct copy of the DNA it will contain both the coding (exons) and the non-coding sequences (introns) of the gene. Introns are removed from this pre-messenger RNA and the subsequent molecule is the final mRNA. The mRNA molecules are transported from the nucleus into the cytoplasm where the message is translated into a sequence of amino acids by rRNA in bodies known as ribosomes. Amino acids are brought to the ribosomes by tRNA molecules, each specifying a particular amino acid, and synthesised, in the presence of rRNA, into a linear sequence.

Genetic Code

The genetic code is the set of rules by which information encoded in genetic material is translated into proteins (amino acid sequences) by living cells. The code defines how sequences of three nucleotides, called codons, specify which amino acid will be added next during protein synthesis. With some exceptions, a three-nucleotide codon in a nucleic acid sequence specifies a single amino acid. Because the vast majority of genes are encoded with exactly the same code, this particular code is often referred to as the canonical or standard genetic code, or simply the genetic code, though in fact there are many variant odes. For example, protein synthesis in human mitochondria s on a genetic code that differs from the standard genetic code. Not all genetic information is stored using the genetic code.

Table 1. The genetic code showing the amino acids coded by the 64 triplet combinations of the four bases. The bases down the left hand side represent the first position in the reading frame, the bases along the top indicate the second position and the bases down the right-hand side show the third position

1st base	2nd base				3rd base
	U	C	A	G	
U	Phe	Ser	Tyr	Cys	U
	Phe	Ser	Tyr	Cys	C
	Leu	Ser	Stop	Stop	A
	Leu	Ser	Stop	Trp	G
C	Leu	Pro	His	Arg	U
	Leu	Pro	His	Arg	C
	Leu	Pro	Gln	Arg	A
	Leu	Pro	Gln	Arg	G
A	Ile	Thr	Asn	Ser	U
	Ile	Thr	Asn	Ser	C
	Ile	Thr	Lys	Arg	A
	Met	Thr	Lys	Arg	G
G	Val	Ala	Asp	Gly	U
	Val	Ala	Asp	Gly	C
	Val	Ala	Glu	Gly	A
	Val	Ala	Glu	Gly	G

Abbreviations for amino acids: Alanine (Ala), Arginine (Arg), Asparagine (Asn), Aspartic acid (Asp), Cysteine (Cys), Glutamic acid (Glu), Glutamine (Gln), Glycine (Gly), Histidine (His), Isoleucine (Ile), Leucine (Leu), Lysine (Lys), Methionine (Met), Phenylanaline (Phe), Proline (Pro), Serine (Ser), Threonine (Thr), Tryptophan (Trp), Tyrosine (Tyr), Valine (Val).

All organisms' DNA contains regulatory sequences, intergenic segments, chromosomal structural areas, and other non-coding DNA that can contribute greatly to phenotype. Those elements operate under sets of rules that are distinct from the codon-to-amino acid paradigm underlying the genetic code. The 'words' of the code consist of three bases. There are $4^3 = 64$ possible combinations of the four bases into a triplet code and it is these 64 triplet codons which define the 20 amino acids. Because there are more than 20 codons, the genetic code has some redundancy—most amino acids are coded for by more than one codon.

The codons are written using the symbol U, for uracil (in mRNA), rather than T, for thymine (in DNA). Three codons (UAA, UAG and UGA) do not encode amino acids but act as signals for protein synthesis to stop and are called termination codons or stop codons. The triplet AUG codes for methionine (formyl methionine in bacteria and mitochondria) and is the signal for protein synthesis to start. It is thus the initiation codon which sets the reading frame. The amino acid sequence of all proteins therefore starts with methionine but this is sometimes removed later.

Protein Structure

Proteins have many tasks. Some form the structure of tissues, others—the enzymes -act as extremely specific catalysts of biochemical reactions, and yet other proteins, such as hormones, have a regulatory function. By their very nature proteins are bound to be highly complex molecules, but it is possible to categorise their structure into four basic levels. Thus, protein secondary structure is based on characteristic patterns produced by the properties and interactions of particular types of amino acids within the chain. One such secondary structure is an alpha-helix, another is a pleated sheet. The tertiary structure is dependent on how these secondary structures become folded in three dimensions. Therefore the DNA code, through the linear relationship of the various amino acids, dictates both the secondary and tertiary structures. This is an important point because it reveals that point mutations in the DNA coding for a particular protein can have far-reaching consequences on the final size, shape and overall charge of that protein.

Many proteins are composed of two or more polypeptide chains (subunits) and the subunits making up a protein may be identical or they may be different. The generic name for proteins with more than a single subunit is oligomers. This is the level of quaternary structure of proteins and it enables larger proteins to be produced without requiring a very long gene sequence in the DNA. It also allows greater functionality in proteins by combining different activities within a single molecule. Proteins with a single subunit are called monomers, those with two subunits are dimers and those with four are tetramers. For example glucose-phosphate-isomerase, an enzyme involved in the production of energy from the breakdown of carbohydrates, is a dimer, with two subunits coded by the same gene, while haemoglobin, which carries oxygen around in the blood, is a tetrameric molecule consisting of two alpha-globin and two beta-globin chains each coded by different genes.

Chromosomes

In fish and shellfish, as with all other eukaryote organisms, the DNA molecules in the nucleus are combined with proteins, mainly histones, to make chromosomes. Each chromosome represents a single DNA molecule. Chromosomes are usually only clearly visible and identifiable when cells are dividing, at which time the chromosomes have already divided into daughter chromatids. However, the daughter chromatids retain connection to each other at a position called the centromere, or primary constriction, and this is the last part of the chromosome to divide. The position of the centromere on the chromosome can be central (metacentric), between the centre and one end (submetacentric), very close to one end (acrocentric), or terminal (telocentric). The number of chromosomes, their lengths and the positions of their centromeres are unique to each species and these characters are used as descriptors for the species karyotype. Chromosomes themselves mutate and evolve and before the advent of allozyme markers some geneticists spent much of their time squinting down microscopes following the inheritance of chromosomal rearrangements. Nowadays, chromosomal variation is assessed for aquaculture and fisheries purposes, mainly in relation to interspecies hybridisations.

Almost all fish and shellfish are diploids. Diploids have two complete sets of DNA instructions, so that each chromosome is just one of a homologous pair. Normal cell division—mitosis—combines division with replication because prior to cell division each chromosome replicates (into two chromatids) and one copy passes into each daughter cell. Diploidy is thus maintained. During the process of sexual reproduction a specialised cell division called meiosis takes place which produces daughter cells, the gametes, which have only a single set of chromosomes. Gametes are therefore haploid. If an organism inherits the same version of a gene from both parents, it is said to be homozygous. If the two versions are different, the organism is heterozygous. Each version of a particular gene is called an allele; the two alleles possessed by a diploid organism at each locus (position on the chromosome, plural loci) make up its genotype for that locus.

In many organisms there is a special pair of chromosomes which defines the sex of their carriers. For example in the XX-XY system present in humans and some fish, females have a pair of identical sex chromosomes (the X chromosomes) while males have one X chromosome and a reduced size Y chromosome. The other chromosome pairs are called autosomes. In many shellfish there are no identifiable sex chromosomes and, in the case of certain molluscs such as oysters, individuals may even change their sex during their lives.

Sexual Reproduction

The key feature of sexual reproduction is the production of haploid gametes through the process of meiosis and the uniting of these gametes to produce a new diploid generation. The process of meiosis shuffles the genetic material in such a way that none of the haploid chromosome sets in the gametes are identical to either of the haploid sets present in the parent from which they are derived. The process of meiosis actually consists of two cell divisions, meiosis I and meiosis II. Meiosis I begins long before the chromosomes become clearly visible. The chromosomes are initially very thin and uncontracted but become progressively more contracted and more visible during the prophase stage.

During this stage, homologous pairs of chromosomes come to lie closely together and at the same time each chromosome in each pair divides into chromatids that remain attached to one another at the centromere. So each pair of chromosomes consists of four chromatids. Such pairs of chromosomes at this stage are called bivalents. It is during this time, while the chromosome pairs are adhered closely together, that the process of recombination or crossing-over occurs. Recombination involves the interaction between two ordinary (double-stranded) DNA molecules. The effect is that both molecules break, but the ends rejoin to the 'wrong' molecule. From the genetic point of view, each chromatid in a bivalent can be considered to be effectively a single DNA molecule and the recombination interaction takes place between chromatids that derive from different chromosomes of the pair—the non-sister chromatids. These recombinations take place in every chromosome pair, usually one per chromosome arm, but sometimes more than one. The place where a recombination event is located on a bivalent is visible as a hiasma.

At metaphase of meiosis I the bivalents lie across the equator of the spindle with their centromeres attached to the arms of the spindle. Meiosis I is a reduction division. During anaphase, each pair of chromosomes is separated so that one of the pair goes into one daughter cell and the other into the other daughter cell. So at the start of meiosis I the cell contains four copies of the genetic information, but after division each daughter cell contains only two copies of the genetic material. Meiosis II is effectively a mitotic division where the two chromatids from each chromosome separate into daughter cells. Therefore each diploid cell that enters into meiosis produces four haploid gametes. Some genetic texts suggest that a cell can be regarded as 'tetraploid' when it enters meiosis I because it has four copies of the DNA.

What is the actual effect of recombination across the genome? Take a species such as the flat oyster *Ostrea edulis,* for example, which has 10 pairs of chromosomes. In all cells of the body, including the germ cells which will undergo meiosis, one of each pair of chromosomes will have come from the female and the other from the male parent. Envisage one of the chromosomes as a linear arrangement of genes along a single molecule of DNA. The other chromosome of the homologous pair will also consist of that same linear arrangement of genes along its length but will have come from a different parent. It has the same genes, in the same order, but has a different ancestry. That ancestry will have provided it with different variations at many of its genes compared with the other chromosome of the pair. In early meiosis each chromosome has replicated itself so there are two DNA copies (chromatids) of each chromosome. Recombination occurs between non-sister chromatids such that a stretch of DNA from one chromatid becomes exchanged for the equivalent stretch from the other chromatid. The resulting chromatid DNA molecules that have undergone recombination are therefore different from either of the parental ones. Any chromatids which have not been involved in a recombination event, of course, remain unaltered.

Now note that this process is taking place in all of the 10 pairs of chromosomes in that germ cell during that division. Then consider that this is just one germ cell among the millions of germ cells in the gonad of the oyster. All the other germ cells are also undergoing a meiotic division during which recombination is taking place in all the pairs of chromosomes. The precise position along the DNA molecules (chromatids) at which recombination events take place is (to some extent) random and will generally be different in each dividing germ cell. So it is easy to understand why the 10 DNA molecules (chromosomes) in an oyster gamete are going to be different from any of the 10 parental DNA molecules (chromosomes) that were present in the germ cell before meiosis. It can also be understood why the genetic make up (the 10 DNA molecules) of every gamete is likely to be different from every other gamete.

It can be seen that very extensive shuffling of the genome is achieved by recombination. However, this is not the only reshuffling that takes place during meiosis. Consider the 10 pairs of chromosomes in the germ cell, with one from each pair derived from the male parent of the oyster, and the other from the female parent. These can be indicated as M1, M2, M3 up to M10 (for the male derived chromosomes) and F1, F2, F3 up to

F10. Each of the daughter cells following meiosis I will contain 10 chromosomes, but they will be a random mixture of F and M chromosomes. For 10 chromosomes there are $2^{10} = 1024$ possible combinations. This is called independent assortment of chromosomes.

Therefore, shuffling of the genome takes place by two processes in meiosis -recombination and independent assortment—and these processes probably ensure that no two gametes are ever likely to be identical to either parental chromosome set, nor to one another.

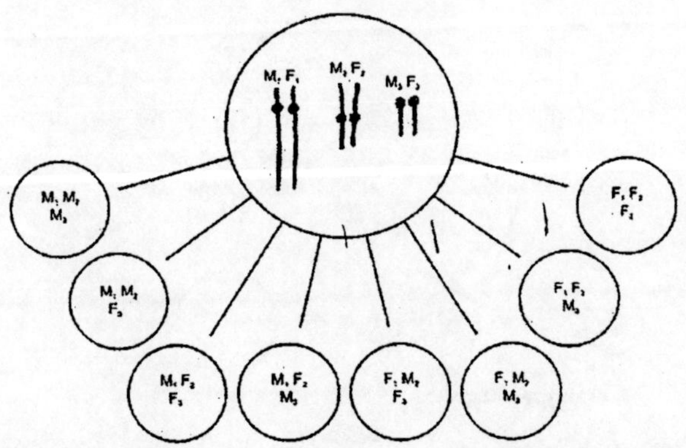

Fig. 4. How independent assortment of chromosomes at the end of meiosis I creates variation. Three pairs of chromosomes are indicated: M1, M2 and M3 are the male parental chromosomes and F1, F2 and F3 are the female parental chromosomes. Independent assortment gives eight possible combinations of chromosomes in daughter cells.

The final part of the process of sexual reproduction, syngamy—the fusion of male and female gametes at fertilisation to form a zygote—further increases the genetic variation of offspring from their parents. Considering all these factors, it is not at all surprising that no two individuals in a sexually reproducing species are identical. The only exception is if an already fertilised egg (a zygote) divides to produce two separate cells, both of which develop independently into normal embryos.

These are monozygotic twins and are effectively genetic clones of one another. The original zygote from which they arose would still have been different from any other zygote, or individual, in that species.

Although the chromosomal behaviour during meiosis is the same in both males and females, there is an important difference in the production of spermatozoa and eggs (ova). In males, four spermatozoa are produced from each germ cell. In females, only one egg (ovum) is produced from each germ cell or primary oocyte (Fig. 5).

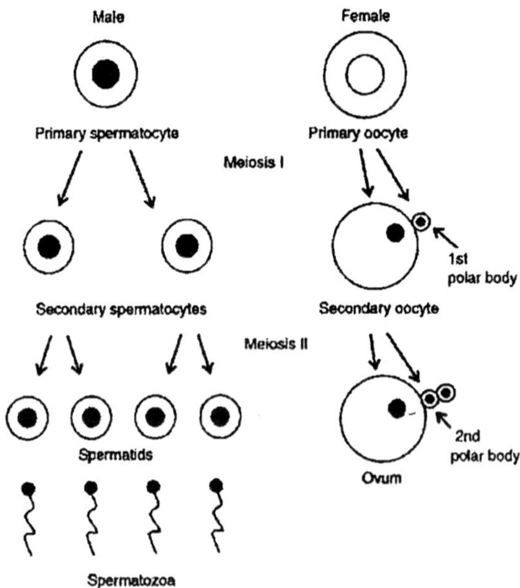

Fig. 5. The difference in meiotic products in males and females. In males each primary spermatocyte produces four spermatozoa while in females each primary oocyte produces a single ovum.

One of the two cells produced during meiosis I is large, and the other is very small, so small that effectively it is really just the chromosomal material with little or no cytoplasm. This small cell—the first polar body—usually does not undergo meiosis II. The large cell—the secondary oocyte—again divides unequally during meiosis II to produce the large ovum and the small second polar body.

Agricultural animals are mostly mammals or birds in which the meiotic divisions take place inside the body of the animal or inside a shell and so are not easily accessible. However, in most fish species, only meiosis I takes place before spawning and the secondary oocytes are released into the water. Meiosis II only occurs when spermatozoa have become attached. In bivalve molluscan shellfish, the oocytes are spawned at metaphase of meiosis I and further development is dependent on the attachment of spermatozoa. This feature of fish and certain shellfish enables chromosome set manipulation to be simply engineered in such species for the production of polyploids. In addition to variation at the DNA level, and the shuffling of the genes during meiosis, genetic variation also occurs at the level of the chromosomes. Such variations are not very useful as genetic markers in aquaculture species, but chromosomal rearrangements have played an important role in evolution.

Mitochondrial DNA

There is actually more DNA in the cell—extra-chromosomal genes, contained within energy-generating organelles, mitochondria, of which there may be several hundred in each cell. In plants, the photosynthetic organelles—chloroplasts -also contain DNA. Animal mitochondrial (mt) DNA is normally present as a circular molecule of around 16 kb in length and there are around 10 copies of the DNA in each mitochondrion in humans. Unlike the chromosomal DNA, there is no meiosis and replication appears to be a simple copying process, though the very latest research does point to there being some form of recombination during mtDNA replication.

Because there are large numbers of mitochondria in an egg, but very few in a spermatozoon, it is hardly surprising to find that the mtDNA present in a sexually reproduced offspring is usually inherited entirely from its mother. This maternal-only inheritance of mtDNA is the normal situation in almost all animals. However, one exception to this rule occurs in an important aquaculture species, the mussel *Mytilus* spp., which has a form of bi-parental inheritance of mtDNA. Females have an F type of mtDNA in every body cell while males have both the F type and an M type mtDNA in most cells of the body. The M type is highly concentrated in the male gonad and is thought to be the only mtDNA present in the spermatozoa. These M mtDNA molecules present in a spermatozoon enter the egg and, in some way which is not yet fully understood, the M type remains in the egg after fertilisation and is eliminated in individuals destined to become female, but retained and preferentially replicated in individuals destined to become males. This unusual arrangement has been named 'doubly uniparental inheritance' (DUI). DUI has recently been detected in other bivalves besides mussels and may be more widespread than currently thought. In contrast to the nuclear genome, the mitochondrial genes of animals are very efficient and have no introns. In addition there is virtually no 'junk DNA or repetitive sequences in the mitochondrial genome, although the control region does often vary in length due to tandem repeats. Exceptions to this general rule are the scallops, many species of which exhibit several large (up to 1.4 kb) repeated sequences within the mtDNA genome which can consequently extend to beyond 30 kb in length.

Measurement of Genetic Variations

Genetic variation can be measured and quantified at several levels. First, the precise sequence of a length of DNA, and how it varies between individuals, can be determined. Secondly, differences between sizes of DNA fragments can be identified.

DNA Sequence Variation

The crude extraction of DNA from animal or plant tissue is a simple process which involves mechanically or chemically breaking down the insoluble cellular structures and removing them by centrifugation. Soluble cellular proteins, and the proteins which bind the DNA into the chromosomes, can then be broken down using a strong protease enzyme and removed, usually using solvents such as phenol-chloroform. The DNA is present in the water-soluble component and can then be precipitated using an alcohol. There are a number of commercial kits on the market which enable further purification of DNA. The next problem is to produce multiple copies of specific fragments of DNA and this can be done either by cloning the fragment or by the use of the polymerase chain reaction (PCR). In the process of cloning, the target DNA is inserted into a vector molecule which is taken up or inserted into host cells. Subsequent rapid replication of these host cells and the vector molecules inside them results in the production of millions of copies of the target DNA. As far as most DNA markers are concerned, cloning is usually only needed during the development phase—once the DNA sequences flanking the markers have been found from the cloned fragments, PCR can be used to produce millions of copies of the target sequence within a few hours. The PCR method relies on the fact that double-stranded DNA becomes denatured and separates into single strands when heated above 90°C. Once denatured, the temperature is lowered to a predetermined annealing temperature which allows short manufactured lengths of single-stranded DNA of known sequence (primers), designed to be complementary to the regions flanking the target DNA, to attach (anneal) to these flanking regions. Raising the temperature to 72°C in the presence of a DNA polymerase enzyme and the building blocks of DNA results in two copies of the double-stranded target DNA.

The sizes of pieces of DNA produced from cloning or PCR can be determined by subjecting the DNA to electrophoresis alongside known size-standards. Since electrophoresis separates fragments based on their sizes, it can be used to purify DNA fragments. For example, the results of a PCR reaction can be run (elec-trophoresed) on an agarose gel. The DNA is stained during or after electrophoresis with ethidium bromide which fluoresces under UV light. Comparison of the sequence between individuals, between populations, between species or between higher order systematic divisions, provides information about the relatedness between these categories. Of course, different classes of DNA are needed to address these different levels of relatedness.

Let first consider a mutation within the coded part (exon) of a gene that codes for an enzyme. However, the mutation could occur at the third base of a codon and, because of the redundancy of the genetic code, will be unlikely to change the amino acid coded for. Alternatively, it could change one of the amino acids in the enzyme produced, but even this may not have any effect on the ability of the enzyme to carry out its cellular biochemical function. Nevertheless, *some* mutations within the exon of an enzyme gene are bound to have a deleterious effect such that individuals carrying that mutation produce an ineffective enzyme and are less likely to survive. Exceptionally, a mutation might be advantageous and improve performance of an enzyme. So enzyme exon DNA sequences are free to change slowly over evolutionary time, at a rate that is considerably less than the rate of mutation, and the rate varies between different enzymes depending partly on the specificity of their biochemical task in the cell.

Finally, let consider sequences that code not for proteins, but for the very RNA molecules which are involved in the process of translation of the DNA code. Here, almost every letter of the code is critical to the functioning of the RNA product and almost any mutation will render it non-functional. The strongly deleterious effect on any individual subjected to such a mutation means that the rate of evolutionary change of these parts of the DNA molecule is extremely slow.

Restriction Fragment Length Polymorphisms (RFLPs)

We can make good use of fragments of DNA as genetic markers without going through the procedure of sequencing them. The various lengths produced can be separated by size and stained on an agarose gel. The same piece of DNA from different individuals will produce different sets of restricted fragments if there have been point mutations affecting the RE recognition sequences. In this way, polymorphisms can be identified based on the pattern of the size fragments on the agarose gel.

Variable Number Tandem Repeats (VNTR)

Variation in the sequence of DNA can occur at certain sites by a method which is not point mutation. Spread throughout the genome are regions called variable number tandem repeats (VNTR), also known as simple tandem repeats (STR) or simple sequence length polymorphisms (SSLPs), which contain tandem (i.e. linked in chains) repeats of DNA sequences. The sequences may be very short (from 1 to 10 bp) or much longer, but the key feature of these tandem repeats is that the number of repeats can vary between individuals. It is thought that increases or decreases in the number of the repeats occur during copying by recombination or replication slippage and that these processes are not only independent of point mutations, but also occur at a much faster rates. Variation in the number of repeats at these satellite (repeated units 100 to 5000 bp), minisatellite (repeated units 5 to 100 bp) or microsatellite (repeated units 2 to 4 bp) loci can be very extensive in populations and provides a valuable tool for investigation of population genetic changes in the recent past. Microsatellite markers in particular are now used extensively for a number of reasons: because they are co-dominant (both alleles can be identified) and therefore can be analysed under the standard Hardy-Weinberg model; because, as 'junk DNA they can usually be considered to be free of selective pressures; because of the high number of both loci and alleles at each locus; and, not least, because automatic DNA sequencers can be used for automated genotyping at microsatellite loci, vastly increasing the rate at which samples can be processed.

DNA Fingerprinting

DNA fingerprinting can be thought of as a combination of RFLP and VNTR. First, genomic DNA is cut with a particular suite of restriction enzymes and differently sized fragments are separated by electrophoresis. The DNA is transferred from the fragile gel to a nylon membrane by the technique known as Southern blotting and the membrane is then probed with a particular satellite repeat sequence which is common throughout the genome. Fragments that contain the repeat show up as a number of discrete bands after autoradiography. These banding patterns are so variable as to be in practice unique to each individual (the chances of a match between unrelated individuals are millions to one). Since the band[illegible] inherited in a predictable fashion, DNA fingerprinting is a very accurate way of determining parentage and the forensic uses of this method are now well known.

Random Amplified Polymorphic DNA (RAPD)

The RAPD method is based on the principle that the shorter the length of the primers which are used in PCR, the greater is the chance that non-target sequences will be amplified. Using a single 10-mer oligonucleotide as the sole primer, PCR is conducted on raw DNA and the resulting fragments, which come from annealing of the primers all across the genome, are separated on agarose gel. Variations between individuals in the presence or absence of bands, reflect mutational difference at the primer sites. RAPDs suffer from the important criticism (among others) that they are not entirely reliable and repeatable.

Amplified Fragment Length Polymorphism (AFLP)

A method which amplifies randomly selected fragments of the genome much more reliably than RAPD is the AFLP method. AFLP is almost an inverse form of RFLP—the genomic DNA is cut into fragments with restriction enzymes and then just a few of those fragments are selectively amplified using special radiolabelled PCR primers. The resulting fragments are then visualised by separation on a poly-acrylamide gel followed by autoradiography. The products of AFLP—a series of bands of different sizes—are similar to the products of RAPDs and variation between individuals is based on the presence or absence of bands. Variation is the result of point mutation differences in the DNA sequence at the PCR primer sites.

Protein Variation

So far have considered genetic variation at the level of the DNA. However, DNA sequence variation, when transcribed, can give rise to differences in the resulting proteins. It is at this level that genetic variation begins to interact with the environment to affect the survivorship and reproduction of organisms and their genes.

Genetic variation at the level of proteins can be identified and quantified using electrophoresis to separate the different protein products of alleles followed by staining to visualise these protein products. It is possible to stain the gel to display all proteins, but it is more useful to take advantage of the substrate-specific catalytic abilities of the class of proteins known as enzymes. This involves using the specific substrate of an enzyme in a stain overlaid on the gel that will change colour where the substrate is altered by the enzyme. The position of any enzyme variants can therefore be located on the gel. These genetic variants of enzymes are known as allozymes and methods for detecting allozyme variation were first developed in the 1960s. Although nowadays regarded by some as an outdated method, the extensive allozyme data sets produced in the last 40 years of the twentieth century fundamentally shifted the ground upon which geneticists tread.

Although little used in fisheries or aquaculture research, the reader should be made aware of the technique of immunological testing which assesses the relationship between proteins on the basis of the relative strength of the antigen-antibody reaction that they will produce. The high value of VNTR methods in parentage and within species analysis is clear, but the extent to which allozyme data can be informative is also emphasised. Although RAPDs score quite highly for a number of approaches, there are real problems of repeatability and reliability with this method. AFLPs are now the recommended quick screening method for identifying quantitative trait loci.

Phenotypic Variation

There are very few examples of easily identifiable phenotypic variation in aquatic organisms controlled by single genes, or even pairs of genes. The best examples are found in the colouring of ornamental fish. It is interesting to consider just how lucky Gregor Mendel (the nineteenth-century discoverer of the method of genetic inheritance) was to have chosen to work with peas, which had a number of easily identifiable characters (round or wrinkled seeds; tall or short plants) each controlled by single genes. Such easily identifiable single-gene phenotypes are rare in most organisms. Because of the extensive development of DNA and allozyme technologies, the search for single-gene phenotypic variation is now uncommon. However, it is important to realise that the visual identification of varieties can be of critical importance to fish farmers without access to a modern genetic laboratory.

References

Avise, J.C. (1994) *Molecular Markers, Natural History and Evolution.* Chapman & Hall, London.

Brown, T.A. (1999) *Genomes.* Bios Scientific Publishers, Oxford.

Carvalho, G.R. & Pitcher, T.J. (eds) (1995) *Molecular Genetics in Fisheries.* Chapman & Hall, London

Majerus, M., Amos, W. & Hurst, G. (1996) *Evolution: the Four Billion Year War.* Longman, New York.

Turner, P.C., McClennan, A.G., Bates, A.D. & White, M.R.H. (1997) *Instant Notes in Molecular Biology.* Bios Scientific Publishers, Oxford.

Winter, P.C., Hickey, G.I. & Fletcher, H.L. (1998) *Instant Notes in Genetics.* Bios Scientific Publishers, Oxford.

3

Fish Population Genetics and Fisheries Management

Of all the animals and plants in the aquatic environment fish is the most important source of human food. They are the major source of protein for many people and constitute the main part of the diet in many cultures. Fish and fisheries not only provide a significant portion of the protein available for human consumption, but they are also an economically significant activity, providing jobs and investment opportunities and, for many countries, a means of improving the balance of international trade.

Most of the fish used for human consumption is obtained through exploitation of wild populations. The management of the wild populations comprising commercial or sport fisheries presents genetic problems that are unique to fisheries management. Reduction in the genetic resources of natural fish populations has become an important fisheries management problem. Much of the reduction is due to various human activities. Not only has the genetic diversity of many fish populations been altered, but many thousands of populations and species have been extirpated by pollution, over fishing exploitation, destruction of habitat, blockage of migration routes and other human developments. In nearly all cases, fishery management has largely been concerned with the immediate resource of interest; that is, the abundance and size of fish available for harvesting. This short-term focus may be economically advantageous in the short run, but in the long term may cause extinction of the population. Concern with reduction of genetic resources in fish is part of a larger global concern for the genetic resources of the biosphere. For this reason, molecular genetic research should be strongly supported, for it is vital to the long-term management of fisheries resources. This approach addresses two slightly different aspects of genetic resources: conservation of gene pools and conservation of genetic diversity. The study of the genetic variation in populations and its change, the following of allele frequencies in populations through time and space, is the main subject of population genetics. Genetic variation is the raw material in a species and populations, which enables them to adapt to changes in their environment. New genetic variation arises in a population from either spontaneous mutation of a gene or by immigration from a population of genetically different individuals. The number and relative abundance of alleles in a population is a measure of genetic variation.

Genetics and fishery management can interact in several ways. When the genetic population structure of a species is known, the distribution of subpopulations in mixed fisheries can be estimated . Regulation of harvest to protect weaker populations can be made based on these distributions. It is important to identify and regulate for genetic changes within a population because of differential harvests because of the drastic and long-term effects they may have on a population.

The genetic study of natural populations is dependent on the availability of polymorphic neutral markers. Although electrophoresis of proteins has been widely used for the direct study of genetic variation in fish populations, DNA markers are becoming more popular in order to obtain information about gene flow, allele frequencies and other parameters that are crucial in population biology . Biologically important characteristics of populations, including their size and productive efficiency, are determined by the historically established gene pools. Therefore, the population genetic analysis of species in nature is of primary importance in developing an optimal strategy for their effective management. Such a strategy should provide not only for maximum economic benefits but also for continuous maintenance of natural populations. Fisheries biologists must emphasize the importance of elucidating the factors and conditions that permit populations and species to be maintained.

The Stock Concept

Worldwide, fisheries provide a source of employment for millions of people and a source of protein for millions more. The management of fisheries is therefore a critical activity. Unfortunately, the realisation that fish are a limited resource was slow to reach acceptance and there is not much evidence that attempts to manage fisheries have led to sustainability.

In order to manage a fishery effectively, a great deal of information is required. Most of these factors can be, and are, determined from fishery statistics based on detailed surveys and analyses of landings. However, a critical requirement is to know whether the fish species exists as a single genetic unit or as a series of relatively genetically distinct groups. Is the species genetically homogeneous or is it genetically heterogeneous? If there are local genetically distinct groups, management strategies will need to be adapted to take this into account. The substructuring of fish into different, relatively genetically discrete groups may take various forms. For example, the migratory salmonids return to the rivers in which their parents spawned and this effectively creates genetic groups that are associated with particular rivers or even particular regions of a river. In another example, some lake fish have autumn and spring spawners and as long as this character is inherited, two genetically distinct groups can evolve. In many marine fish, there are particular sites where spawning aggregations occur and these can lead to genetic differentiation. In the case of sedentary shellfish, dispersal is usually via a larval phase and hydro-graphic factors will constrain the direction and distance of dispersal. Shellfish such as flat oysters with a short larval phase or direct development will be more likely to be genetically differentiated than shellfish such as mussels with an extended larval phase.

All of these terms have been used in the past to describe genetic differentiation within species but they have often not been clearly defined. In the case of fisheries, the word 'stock' has often been defined by fishery managers as a group of fish exploited by a specific method, or existing in a particular area. Although this may be convenient for the analysis of landings or catch and effort data, and for the enforcement of management measures such as quotas or net size, it may bear little relation to the true genetic substructuring of the fish species. The important point to realise is that there is a continuum of genetic differentiation. In some species (or in parts of their range) there may be clearly genetically distinct groups, while in others (or in other parts of their range) there will be groups which are virtually indistinguishable genetically. Thus, the use of a single word such as 'stock' to define the units of such substructuring is not often appropriate.

Therefore, although there will be the same range of genotypes in an offspring generation as in a parent generation, the two are not directly linked by inheritance. What is linked by inheritance, however, is the presence of an allele and the frequency of that allele in the population. And it is the allele frequency (or gene frequency) that is used to investigate the structure of populations within species.

Estimation of Allele Frequencies

When genotypes at a co-dominant locus are scored from a sample of individuals, the frequencies of the alleles at the locus can easily be calculated from these data. Remember that every individual has pairs of chromosomes, so they must also have pairs of alleles at each locus, that a homozygote at a locus has two of the same allele, and a heterozygote has two different alleles. Although a diploid individual can only exhibit two alleles at a locus, there may be many different alleles at that locus present in the population as a whole. Allele frequencies are the critical starting point for all further analyses of genetics in populations. By convention, the frequencies of alleles at a locus are symbolised using the lower-case letters *p, q, r, s* and the frequency of an allele is given by:

$$p = \frac{2H_0 + H_C}{2N}$$

where H_o = number of homozygotes for that allele, H_e = number of heterozygotes for the allele and N = number of individuals scored at the locus.

Relationship between Alleles and Genotypes

Considering a single locus, it is convenient to envisage a population as a 'pool' of alleles. Haploid gametes are produced each generation and these gametes each contain a single allele from this pool. Let assume that all adults in the population provide equal numbers of gametes and that each gamete has an equal chance of combining with all other gametes. This principle was first quantified in 1908 by both Godfrey Hardy, a British mathematician, and Wilhelm Weinberg, a German doctor, and is called, unsurprisingly, the Hardy-Weinberg model. It can be seen from Figure 1 that with random mixing of eggs and sperm the proportions of the genotypes

will be:

AA AB BB

p^2 $2pq$ q^2

which, in mathematical terms, is equal to $(p + q)^2$. As long as nothing upsets the Hardy-Weinberg model, allele frequencies should remain constant from generation to generation. The frequencies of the genotypes predicted by the model are dependent on the assumptions of equal and random parental contributions and equal and random mixing of gametes.

Eggs / Sperm	p	q
p	$p.p$	$p.q$
q	$p.q$	$q.q$

Fig. 1. Illustration of the principle behind the Hardy-Weinberg model. Here consider the frequencies of genotypes produced by random combination of eggs and spermatozoa at a locus with two different alleles, A and B, at frequencies p and q.

However, if this is not the case, the model will not hold true. Also, do not usually sample animals at the beginning of their lives; normally fish or shellfish are sampled as adults. In order for the Hardy-Weinberg model to be appropriate in populations of adults, further assumptions need to be made. Assume that there is no selection that causes one or other genotype to suffer differential mortality. So migration and selection are both factors that can cause deviation from the genotype frequencies predicted by the Hardy-Weinberg model. A further consideration is that very few fish or shellfish species exist as discrete generations, most populations being mixtures of animals from different overlapping generations.

Change in Allele Frequencies

If a population is in agreement with the Hardy-Weinberg model then allele frequencies at a locus are not being driven (by selection or migration) to change significantly from generation to generation. However, there is always bound to be some variation in allele frequency from one generation to the next because of the element of chance. Contrary to the idealised situation required to make mathematical predictions, in real life all individuals of all genotypes will not produce exactly the same number of gametes. Neither are all gametes likely to undergo truly random mixing. Therefore there will be variation from the model in every generation. This is a fact of biological variability, called random genetic drift. In order to quantify this natural variation, allele frequencies can be given a variance as indicated in the formula:

$$\text{variance of frequency of allele} = p = \frac{(1-p)}{2N_c}$$

where p is the frequency of the allele and N_c is the effective population size. The concept of the effective population size is an important one. It is the number of individuals in the population that contribute genetically to the next generation. It will exclude juveniles, individuals too old to reproduce and those that provide non-contributory gametes These may be individuals that have infertile gametes, or those whose gametes never come into contact with the gametes of other members of the population due to geographic position or timing of spawning. Thus it can be seen how the effective population size could, in certain circumstances, be very much smaller than the total number of individuals in the population.

If the effective population size (N_e) in the expression above is a very large number, then the allele frequency variance will be very small. Conversely, allele frequency variance will be large when N_c is small. This means that natural fluctuations in allele frequency between generations will be much greater in small populations. Allele frequencies in a population can therefore change over time owing to random genetic drift. However, they may also change owing to pressure of selection (one genotype or one allele survives better than others at a locus) or to patterns of migration or dispersal that may fluctuate in direction or in strength over time.

Population Structure

The differentiation of a species into genetically different populations is a fundamental part of the process of evolution. For a number of physical or biological reasons the distribution of a species may become fragmented. For example, during the last ice ages, aquatic species in the temperate regions of the northern hemisphere were driven south and many became restricted to local areas in southern habitats. Where species were fragmented into different areas, many of these populations remained isolated for long periods of time in such refugia. Following the retreat of the last glaciation some 10 000 to 13 000 years ago, aquatic and terrestrial species went through the process of gradual recolonisation of habitat and this has provided plenty of opportunity for further fragmentation. Once fragmentation has occurred, allele frequencies at most loci will be subject to random genetic drift in the fragmented populations. As have seen, random genetic drift produces more-rapid allele frequency change in small, compared with large, populations. If only a few individuals are involved initially in founding a population then allele frequencies in the new population may be very different from the source population, and if numbers remain small further allele frequency changes may be rapid.

In addition to random genetic drift, localised adaptation will occur involving selection at some loci for particular characteristics and these will cause further differences between populations. These processes of random genetic drift and adaptation will tend to increase differences between populations while migration or larval dispersal between populations will tend to reduce it. So current population differentiation in a species is essentially the result of the historical interplay between the environment and the forces of dispersal, adaptation and random genetic drift.

Genetic Markers and Population Structure

In any population study, the ideal first step would be to collect samples of the species across its entire range to estimate genetic differentiation within the species as a whole. In fact, this is seldom done. For reasons of economy most genetic studies related to fisheries have tended to focus on limited sampling in specific areas where there has been a commercial interest. Depending on the type of marker employed, genotypes (allozymes, microsatellites) or haplotypes (mtDNA) are scored for the individuals sampled and the data are analysed in a variety of ways to quantify levels of genetic variation between populations. One commonly-applied type of analysis, called *F*-statistics, was developed by Sewall Wright in the first half of the twentieth century. Because the development of mathematical models requires certain concepts or parameters to be fixed at the start, various assumptions had to be made in the development of *F*-statistics and it is important to consider what these were.

First, Wright made the assumption that all populations were of the same size, that is, they consisted of approximately the same number of reproductively active individuals. So, in this instance, population size is not the same as geographical size. How valid is this assumption? Well, it is easy to imagine situations, particularly in a heavily-fished species, where population size could be very different between populations.

Also, in sedentary shellfish, the size of populations will be constrained by, among other factors, the availability of suitable habitat.

In most cases, movement of larval individuals between populations is likely to be unidirectional rather than multidirectional due to the effects of currents and there are very few fish species where exchanges of juveniles or adult individuals between populations would be expected to be random in both directions. These two theoretical assumptions are a required part of the 'island model' which Wright used as a basis for *F*-statistics. It was further assumed that changes in allele frequency at most loci over time were essentially due to random genetic drift rather than to selection. Wright's *F*-statistics provide answers to two different questions. The first question is: for the loci scored, are the genotypes in the proportions predicted by the Hardy-Weinberg model? F_{IS} provides a measure of this agreement for a single population and F_{IT} for all the populations combined. F_{IS} and F_{IT} can vary from -1.0 through zero to 1.0 and exact agreement to the Hardy-Weinberg model equals zero. The second question is: for the loci scored, are the allele frequencies different between various populations? F_{ST} provides a measure of this population differentiation and ranges from zero, where all populations have the same allele frequencies at all loci, to 1.0 where all populations are fixed for different alleles at all loci.

The Japanese geneticist Masatoshi Nei developed G_{ST}, the coefficient of genetic diversity, which is an equivalent index to Wright's F_{ST}. It is calculated slightly differently, but is essentially addressing the same question. Because these indices represent the genetic differentiation between populations, it is possible to suggest how many individuals might be being exchanged per generation between these populations in order to produce the amount of differentiation observed.

With the exchange of individuals comes the exchange of genes and the greater the exchange, the less will be the genetic differentiation between the populations. Although there is no hard and fast rule, it is accepted that important or significant genetic differentiation is only likely to arise between populations when fewer than one individual per generation, on average, is being exchanged. Again, this is based on the assumption that changes in allele frequencies are based on random genetic drift alone. Obviously, at some loci, there could be significant changes due to selection and such changes would give the false impression that there is less migration between populations than is actually the case.

There are alternative ways of estimating genetic differentiation besides using Wright's F_{ST} or Nei's G_{ST}. It is possible simply to test for the heterogeneity of allele frequencies at each locus across all populations using Contingency Table tests. Traditional testing would involve the use of the χ^2 (chi-squared), or *G*, test, but nowadays, with the enormous computing power available in desktop PCs, much safer 'exact' tests can be employed and these are used in modern genetic analysis computer packages. What this test tells is whether there is significant heterogeneity in allele frequencies across all populations. Selective removal of populations that look particularly different, followed by retesting the data, can reveal further detail. Care is needed to avoid the type I statistical errors associated with the use of several tests of the same hypothesis.

Finally, allele frequencies can be used to calculate pair-wise genetic differences between populations, or species, using formulae developed by a number of scientists. Nei's genetic identity (*I*) and genetic distance (*D*) are the most commonly used. Once again, a major difficulty faced by geneticists in trying to describe what can be an extremely complex situation by a single statistic is the problem of the neutrality of the gene loci used. On the one hand, genetic variation at many loci could be adaptive such that selection operates and particular alleles or genotypes are favoured in particular situations. Alternatively, allelic variation at most gene loci could be neutral, that is, it varies by random genetic drift and is not subject to selection. For most genetic indices, the usual assumption is that all of the alleles at all of the loci included in the determination of the index are neutral.

Genetic Differentiation in Aquatic Organisms

Many studies have been carried out to estimate the amount of genetic subdivision within aquatic species. Table 1 illustrates levels of population substructure in various groups. On average, these levels of G_{ST} (which is equivalent to F_{ST}) are not too far from 0.2, suggesting that, in general, subdivision of populations within species is close to that which allows genetic drift to change allele frequencies in local populations in the face of gene flow by migration.

Table 1. Mean levels of population substructure in aquatic groups based on values of Nei's coefficient of genetic differentiation, G_{ST}, estimated from allozyme data

Taxonomic group	G_{ST} (± standard error)	Number of species
All vertebrates	0.202 ± 0.015	207
Fish	0.135 ± 0.040	79
All invertebrates	0.171 ± 0.020	114
Crustaceans	0.169 ± 0.016	19
Molluscs	0.263 ± 0.036	44

However, these average values hide a wide range of variation in population structure between species. For example, let take two contrasting cases—the salmon, *Salmo salar,* and the mussel, *Mytilus edulis.* The salmon has a life history involving extensive juvenile and adult migration, with mature adults returning to their natal rivers to spawn. Thus, although salmon can be found over a very wide area, they are genetically very much restricted to local populations in their own rivers as this is where the transmission of genes between generations occurs. There are strong barriers to gene flow because very few fish return by mistake to non-natal rivers. Each local population consists of relatively few individuals and the population can be sometimes even further subdivided into year classes that return to spawn in the same year as one another. Values of F_{ST} for salmon, calculated from extensive allozyme data, are in the region of 0.4 and indicate that, as expected, there is very little gene flow ($N_e m$ = 0.38) and strong genetic differentiation into local populations in this species. On the other hand, the mussel is a sedentary organism and adults never actively move more than a very short distance. As with salmon, mussels release eggs and spermatozoa into the water, but, in contrast to salmon, mussel larvae are small and planktonic and are at the mercy of currents that can disperse them over great distances. Larval life, from egg to metamorphosis, in mussels lasts around 4 weeks and can be further extended in two ways. First, if a suitable habitat for settlement is not discovered, metamorphosis can be delayed for more than a week, allowing further dispersal. Secondly, once metamorphosed, these very young mussels (spat) can detach, secrete a very thin byssal thread and be transported in currents by a process known as byssal drifting, which is analogous to the aerial dispersal of spiders by gossamer threads. Typically F_{ST} values from allozyme studies in mussels are very low (<0.001) and clearly reflect extensive gene flow ($N_e m$ > 250) and a lack of any population substructure.

It is also interesting to compare genetic differentiation in marine fish such as Atlantic herring *(Clupeus harengus)* with that in *anadromous* fish such as salmon. The herring has a very large effective population size because all fish return to the same region to spawn and there are therefore few, if any, barriers to gene flow. Average F_{ST} for herring is 0.01 and $N_e m$ is 24.8.

Therefore, the conclusion is that where the effective population size and/or migration is large, gene flow will tend to dominate over random genetic drift, there will be little differentiation and F_{ST} will be close to zero. Where $N_e m$ is small, random genetic drift will tend to dominate over gene flow, allele frequencies will differ strongly between populations and F_{ST} will be large.

Although information about population differentiation is very valuable to fishery managers, it is important to recognise that a *lack of evidence* for local genetic populations does not always mean a *lack of substructure.* If populations are genuinely genetically isolated from one another then would expect random genetic drift to be acting to change allele frequencies at all polymorphic loci. However, for reasons of chance, not all loci will show different allele frequencies in different populations. If the loci screened in limited surveys happen to be these invariant loci then the true variation remains hidden. It is therefore important to examine sufficient loci to be sure that an apparent lack of genetic differentiation can be relied upon.

Allozyme studies suffer from another cause of failure to detect genetic isolation because of their very nature. As enzymes, they are vulnerable to selection operating at the biochemical level, either directly on them or on the biochemical pathways within which they operate. If the selection is such that it is in the same direction in two genetically isolated populations, that is, an allele is at high frequency because the individuals carrying it are favoured in both populations, then the allozyme data will be interpreted as indicating that the populations are not genetically different. On the other hand, two populations with habitat differences but which regularly exchange genes via larval flow may have different allele frequencies at a locus which is under selection where one allele is favoured in one population but is deleterious in the other. Differential mortalities in the two populations will cause allele frequency differences between samples taken from the two populations and this will give a false impression about the degree of reproductive separation of the two.

As an example, show allele frequency data at a protein locus *(PT-A)* for populations of the queen scallop *(Aequipecten (Chlamys) opercularis)* around the British Isles and northern France. There are three alleles at the locus, differing in frequency in different regions. In the south and west, populations are characterised by a high frequency of one allele, another predominates in the north, while the third is the most frequent allele found in the Irish Sea, on the east coast of Britain and in the Bay of St Brieuc. The differences in allele frequencies between the main regions are clearly quite large and the immediate conclusion might be that there are several different populations or 'stocks' of the queen scallop around the British Isles and northern France. However, this begs the question: are these differences really the result of lack of gene flow between populations, allowing the observed genetic differences to have accumulated through genetic drift? Or could there be plenty of larval exchange between populations and the differences be simply the result of selective mortality of certain of the *PT-A* genotypes between the larval and adult stage?

Knowing the life history of these scallops can say that larvae certainly live long enough to allow exchange between these populations, but of course whether this actually happens depends upon the direction and speed of ocean currents. In the case of the Bay of St Brieuc, there is strong evidence of a cyclical gyre that will tend to retain scallop larvae within the confines of the Bay. Also, in the Irish Sea there are a number of important fronts affecting the flow of water which build up at certain times of the year and which could prevent the passage of larvae out of, or into, this body of water.

Accepting the possibility of restricted larval exchange due to hydrographic factors, how long could these populations have been isolated and is it long enough for random genetic drift to have produced such significant differences in allele frequencies? Therefore, this is the maximum time in which genetic drift could have created these differences. Current wisdom, based on Nei's 'molecular clock', is that this is not long enough.

How then do explain this apparent genetic differentiation? Could there be restricted larval exchange together with selection? This is not an easy question to answer. There is no evidence for significant deviations from the Hardy-Weinberg model at this locus in any of the populations studied, so clearly any selection operating is not particularly strong. In order to demonstrate selection, weak or otherwise, would need to know the function of the protein coded for at the *PT-A* locus and link this in some clear way through its biochemistry and physiology to some environmental factor which was different in the different populations. In this case, information on the protein function or any environmental link is not known. Another possibility to consider is that the *PT-A* locus might be located very close on the chromosome to another locus which is under strong selective pressure. In this case the *PT-A* locus might only be a marker for the selected locus.

From the point of view of fisheries management, the safest conclusion that can be drawn from the *PT-A* story is that there are, indeed, significant differences between certain populations of the queen scallop and that they are probably the result of limited larval exchange combined with weak selection for one or other allele in different regions.

Mixed Stock Analysis (MSA)

When an oceanic salmonid catch is brought ashore it probably consists of fish from a mixture of distinct populations from different river sources. If the fishery is managed as a single unit there is the potential for the overexploitation of fish from the less abundant populations, or the suboptimal harvesting of fish from those that are more abundant. In order to enable proper management and obtain the maximum sustainable yield from such fisheries, need to estimate the proportions of the various populations in the fished resource. If the contribution of the various populations varies geographically or over time, then fishing effort can be directed to exploit the strongest populations and pressure on weaker populations can be reduced by targeting fishing to particular regions or particular times.

Early attempts to tag or brand fish as they departed from specific rivers were relatively ineffective because of the high initial labour costs and the loss of markers in a proportion of the fish. Far more effective is the use of genetic markers. Allozyme markers have been widely used in the management of Pacific salmon ,based on comparisons between allele frequency data for the individual river populations and allele frequencies of the total catch. Of course, to make it work, the differences between river populations must be sufficiently great that allele frequencies at, minimally, a few of the loci scored will enable identification of that population in the total catch. Fortunately, there are statistical methods available which make these comparisons possible even when allozyme-allele frequency differences are not very great. The potential resolving power of microsatellite markers is much greater than allozymes and it is expected that they will be used increasingly in MSA.

MSA is also relevant to some sport fisheries that consist of a mixture of natural and introduced stock. The introduced element is either transplanted from another area or is derived from hatchery production. A good example is the trout, *Salmo trutta,* which exists in Europe in two different forms. One, the resident or brown trout, inhabits rivers and lakes and never goes to sea, while the other, the sea trout, is anadromous and, as its name suggests, migrates into the sea and spends some time there before returning to its natal river to spawn. Where waterfalls that are impassable to trout are present in rivers, the populations above the falls will be resident trout and those below will be sea trout. Because sea trout grow bigger than resident trout they make a better fighting fish for rod and line and, mainly for this reason, restocking of river populations above impassable falls is often carried out with hatchery-reared sea trout. The expectation was that there would not be a problem with the survivors of these introduced sea trout breeding with the natural population of resident trout because, once they had migrated to the sea, they would not be able to return to their original position (above impassable falls) in the river to breed. However, allozyme-based studies in various parts of Europe have shown that this does not always happen. It is clear that some stocked sea trout do not migrate to sea but remain in the populations of resident trout with which they will then breed. This leads to a loss of the unique genetic identity of the resident stocks and is an important conservation issue.

Conservation Genetics

Most educated people are now aware of the fragile state of the planet and the increasing pressures from human activities on the animals and plants with which share the biosphere. Species are becoming extinct at a rate comparable to the mass extinctions of geological time and, in addition to loss of species, there is a loss of biodiversity within remaining taxa. In this case, for 'conservation of biodiversity' read 'conservation of genetic diversity'. How then do measure genetic diversity in a species and, having established that genetic diversity is reduced, what can do about it? The commonest measure of genetic diversity is heterozygosity: at a single locus this is the proportion of individuals that are heterozygous and this value can be averaged across as many loci as there are data for. Another measure of diversity is the average number of alleles present across all loci.

One of the difficulties in assessing genetic diversity is that many marine species exhibit the phenomenon of chaotic patchiness. This is the situation where there is extensive microspatial variation in allele frequencies detected at any one sampling time, but allele frequency changes occur over time such that the pattern observed might be very different if sampled at another time. Reproductive strategies in the marine environment are likely to create chaotic patchiness because of the explosive reproductive capacity of just a few individuals. Therefore, the lucky survivors of spawning and larval development are not always the average genetic representatives of the parent population. Fortunately, in the case of commercial marine species -probably because they normally have very large population sizes—studies have tended to show little significant temporal variation in allele frequencies. The problem really arises and becomes acute where population sizes have for some reason (e.g. fishing, habitat loss, disease) become much smaller. When population numbers dwindle to worrying scarcity and the species could become extinct, have to consider how to manage the conservation of the species actively.

Beyond this, there is little that geneticists can do apart from lobbying the appropriate authorities in an attempt to alleviate the pressure on the species. If the situation is so dire that the numbers of individuals in a species fall to very low levels then the problem becomes graver because of the effect of inbreeding depression.

Considering marine fish and shellfish species, complete extinction from overfishing is an unlikely outcome because fishing pressure is usually reduced when it becomes uneconomic to target a particular species. There could still be sufficient numbers of individuals of that species out there to maintain a small but viable population. The type of fish species which are more vulnerable to extinction from overfishing or habitat loss are those which have colonised lakes following the retreat of the ice sheets of the last glaciation. Such species are often strongly substructured, with particular subspecies or variants restricted to just one or a few lakes. It is difficult to secure samples for genetic study from whales. Indeed, until the development of DNA techniques which required only minute quantities of tissue, such as can be obtained from the sloughed-off skin, little genetic information was known about cetaceans. More recently, using mtDNA, microsatellites and DNA fingerprinting, together with observations of behaviour in the wild, much has been learned about the family relationships between individuals within pods. Such information has assisted the conservation effort, not least by revealing and publicising the complicated and subtle nature of the population substructure of these species.

In addition to their value in addressing the wider questions of population structure, molecular genetic markers can be of great forensic importance. Naturally, legislation to protect threatened species or to manage fisheries relies on correct identification of an organism or parts of it. In the case of many fish and shellfish it may often only be a piece of muscle tissue from which a prosecution must be mounted. The large sea scallop *(Placopecten magellanicus)* fishery is of major importance on the Atlantic coast of Canada and the USA but, in the same region and extending across the northern Atlantic, there is a fishery for the smaller Iceland scallop *(Chlamys islandica)*. Scallops for the American market are shucked (removed from the shell) at sea with only the adductor muscle (meat) being landed and a critical management tool for regulation of these fisheries is to use meat counts. Given that the meats look the same, it would be easy for fishers to pass off small, undersized sea scallop meats as Iceland scallop meats.

Although meats are often up to 12 days old by the time they are landed, identification of adductor muscle tissue of the two species has been achieved using two allozyme loci that are diagnostic. In addition, diagnostic markers have been established by RFLP analysis and by direct sequencing of a portion of the 18S rRNA gene. Thus, molecular biology has provided fisheries managers with simple, relatively cheap and effective tools to assist with the enforcement of legislation.

References

Beaumont, A.R. (ed.) (1994) *Genetics and Evolution of Aquatic Organisms.* Chapman & Hall, London.

Hillis, D.M. & Moritz, C. (eds) (1996) *Molecular systematics,* 2nd edn. Sinauer Associates Inc., Sunderland

Ryman, N. & Utter, F. (eds) (1987) *Population Genetics and Fishery Manageme::t.* University of Washington Press, Washington.

Utter, F.M. (1991). Biochemical genetics and fishery management: an historical perspective. *Journal of Fish Biology,* supplement A, 39: 1-20.

Ward, R.D.. Skibinski, D.O.F. & Woodwark, M. (1992) Protein heterozygosity, protein structure and taxonomic differentiation. *Evolutionary Biology,* 26, 73–159.

4

Genetic Interactions in Fish Hatchery

The principal function of a fish hatchery is to produce large numbers of juvenile or market-sized individuals economically. For this purpose a broodstock of adults is initially collected from the wild. This brood-stock may be renewed from the wild each season, or it may be retained and added to, over the years, by selected hatchery-reared individuals. From the genetic perspective, inbreeding increases homozygosity and almost always has deleterious phenotypic effects such as reduced larval and juvenile performance or viability. These phenotypic effects of inbreeding are called inbreeding depression.

Although inbreeding itself will generally reduce the performance of offspring, the process of inbreeding does have a potential value. It can be used to develop highly homozygous 'lines' of aquacultural animals. Such lines will be homozygous at most of their gene loci, but different inbred lines will be homozygous for different alleles at many of the loci. When two separate inbred lines have been produced, these can be crossed with one another to produce 'F1 hybrid' offspring which will then be heterozygous at these loci. F1 hybrids are expected to demonstrate above average performance (hybrid vigour, discussed below) and in agricultural animals and crops this approach has proven to be extremely effective. However, apart from the specific value of F1 hybrids, the first important genetic consideration for the broodstock is to avoid or reduce inbreeding.

Each variant allele at each coding locus in a population can be regarded as part of the 'genetic resource' of that population. An allele alone, or in combination with other alleles or loci, could be responsible for conferring on its carrier a valuable trait such as increased resistance to a particular disease, better cold tolerance or faster growth. Therefore, the loss of any allelic variants is a potential loss of valuable genetic resource. Of course, if most allelic variation at coding loci is neutral, then this is less important, but would be unwise to ignore the certainty that at least *some* variants at coding loci will be advantageous. If not now, then most likely in the near- to medium-term future, global warming will bring about the increasing importance of high temperature resistant allelic variants at biochemically important loci in temperate aquaculture species. Such alleles may be effectively neutral until extreme summer temperatures reveal their value.

Genetic Variation in the Hatchery

In order to meet the demand for oysters, the Pacific or Japanese oyster *(Crassostrea gigas)* was imported from Japan to several areas of the world and this species soon became the most important cultured oyster species worldwide. Several early importations of Pacific oysters were made to British Columbia on the Canadian west coast and the Pacific oyster became naturalised in that region. This naturalised population was a major source of broodstock for hatcheries along the west coast of the USA. In spite of the reduction in abundance of flat oyster *(O. edulis)* in Europe, there remain some natural populations and small-scale hatchery production of this species continues.

The Pacific oyster is amenable to strip-spawning and this method of cutting the oyster open and teasing out the gametes is used in almost all commercial hatcheries. Therefore, there is no uncertainty about the numbers of individuals providing gametes for a mass spawning. On the other hand, there is no guarantee that stripped gametes from all individuals are fully ripe and able to fertilise. This could be one source of the discrepancy between the estimated and the actual number of spawners. Another source of this discrepancy is applicable to other oysters and most other bivalves that are not amenable to strip spawning and thus have to be induced to release their gametes by some chemical or physical stimulus. For practical reasons, this induction is usually carried out with the broodstock held together in a large tank, and the water can become very milky with suspended eggs and sperm from just two or three individuals. All individuals will not necessarily respond to the stimulus and it is virtually impossible to tell whether there are more than just a few individuals spawning. Thus, this can reduce the effective numbers of broodstock. In the case of the flat oysters, which brood their eggs and larvae, control over spawning is even more difficult to achieve.

Once fertilisation has been achieved, it may be that some families perform far better than others. Slow growing families of oyster larvae will be lost from the system because of the size selection that takes place when sieve sizes are increased—only the larger larvae are retained. Alternatively, some families may just have a poor survival history and therefore will not be strongly represented among the hatchery-produced cohort. This is particularly relevant where, as in the example given, each generation of broodstock is used to produce the next, as this is likely to result in inbreeding with concomitant slower growth and higher mortality.

In the case of many commercially important crustacean shellfish there are particular difficulties in the control of reproduction. Simple external fertilisation does not take place—females collect spermatophores from males and then at some later date use sperm from the spermatophores to fertilise the eggs, which they brood on specialised carrying legs. Putting equal numbers of males together in a tank with females is not always successful in ensuring good genetic mixing—in some species, such as *Macrobrachium rosenbergii,* the largest males may succeed in fertilising most of the females and prevent all other males from taking part.

This brings to a separate difficulty: if there is an uneven contribution between males and females, how does this effect the situation? The formula to estimate effective population size in a spawning broodstock where different numbers of males and females are taking part is:

$$N_e = \frac{4\, N_f N_m}{N_f + N_m}$$

where N_e = effective population size, or effective number of spawners, *Nf* = number of females and N_m = number of males. So, if there are 20 individuals of which only three are male, then the effective number of spawners is only 204/20 = 10.2. Therefore, although 20 individuals are spawned, because of the imbalance between males and females, this is effectively equivalent to spawning only about five males and five females. Also, it must be remembered that it is the effective number of spawners that dictates how much of the genetic variation present in the original stock is transmitted into the offspring.

It can be seen that there are a number of explanations as to why, in spite of the best efforts of hatchery managers, the number of broodstock individuals which pass on their genetic variation to hatchery-produced offspring is likely to be fewer than predicted. One practical solution to overcome some of these difficulties is to carry out several separate mini-spawnings between even numbers of males and females, and then combine the resulting embryos. This increases the certainty that the early larvae or fry are derived from as many parents as possible, but can do little to overcome the problems associated with differential growth or survival between families.

Hatcheries and Heterozygosity

So far have considered how genetic variation measured by numbers of alleles can become depleted in hatcheries. Another measure of genetic variation—heterozygosity (the proportion of individuals that are heterozygous at a locus)—might also be expected to decrease. Observed heterozygosity in the hatchery-derived oysters is not significantly lower than heterozygosity in the original populations of either species of oyster.

One explanation for the retention of heterozygosity is that it is not simply the number of alleles at a locus, but rather their relative frequency, that makes the strongest contribution to heterozygosity. Equal allele frequencies at a locus will give the highest heterozygosity and very unequal frequencies the lowest.

It can envisage that very unequal allele frequencies at a particular locus in a source population will tend to produce more equal allele frequencies in a hatchery population for the alleles which are, by chance, retained there. Thus, heterozygosity is not only retained, but possibly increased. However, in the case of loss of an allele at a bi-allelic locus there will be a dramatic loss of heterozygosity to zero. If many of the loci scored are bi-allelic, as is often the case for allozymes in fish and crustaceans, then reductions in numbers of alleles are more likely to be mirrored by reductions in heterozygosity. In fact, where similar studies have been carried out on fish and crustacean shellfish they have indeed generally shown reduced numbers of alleles and reduced heterozygosity. Studies on other bivalve shellfish have been similar to the oyster story, with loss of alleles but little loss of heterozygosity.

Another factor which might account for some retention of heterozygosity in a bivalve hatchery is the common practice of artificial selection of larvae on the basis of size: small, slow-growing larvae are discarded during the process of sieving as pore sizes increase. How might this affect heterozygosity? Geneticists are interested in heterozygosity in individual organisms, as well as in populations. It can ask the question: Is it advantageous for an individual to have two allelic variants expressed at a locus (heterozygosity) rather than having just one allele expressed (homozygosity)? From the point of view of the genome as a whole, the answer would seem to be clear.

Perhaps surprisingly, there is now quite a lot of evidence to support the idea that multiple-locus heterozygosity (MLH) across just a few allozyme loci is positively (albeit weakly) correlated with size in juvenile bivalves. Individuals that are heterozygous at many of the allozyme loci scored in genetic studies are, on average, larger than equivalent individuals of the same age that are homozygous at many of the scored loci. If this correlation is also present during the larval stage, then the loss of small larvae will increase the average heterozygosity of the remaining larval cohort.

Identification to Family Level

The use of a few highly variable microsatellite markers can

make it possible to identify individuals to particular families. With enough variation at three or four loci, the multiple genotypes of individual broodstock parents will be unique to those parents and therefore to each hatchery-cross carried out.

Identification to Population Level

A good illustration of the use of an allozyme marker to detect hatchery-produced individuals among wild stock is provided by the work of Knut Jørstad and colleagues in the development of cod ranching in Norway. Cod is an extremely important commercial species in the North Atlantic and has suffered from overfishing to such an extent that the famous Grand Banks fishery grounds off Newfoundland and Labrador are now completely closed. Some level of artificial production of cod for enhancement of wild stocks was carried out in Norway over much of the twentieth century, but the evaluation of any release project is dependent on a reliable method for the identification of the released fish. Traditional identification methods involve mechanical or chemical tagging, but these can be time-consuming and expensive and they are not always readable in older fish.

Variation at the glucose phosphate isomerase *(GPI-1*)* locus in wild populations of cod off Norway includes the rare *GPI-1* 30* allele. Using electrophoresis of tissue from fin-clips, individuals heterozygous for this rare allele were identified and spawned together. Yolksac larvae, 25% of which would be *GPI-1* 30/30* homozy-gotes, were reared in artificial seawater ponds and later, when mature, these *GPI-1* 30/30* homozygotes were identified from fin-clips and selected for spawning. Offspring of these matings, all *GPI-1* 30/30* homozygotes, were used in a release experiment in Masfjorden together with an equal number of a control group from matings between unselected cod. Juvenile fish from both groups were chemically tagged with oxytetracycline supplied in the diet so that they could be identified when caught the following year.

Table 1 shows the *GPI-1** allele frequencies of the released fish and wild fish of different ages. It is clear that the *GPI-1* 30* allele is rare in all the age classes of wild cod but that it is present at a frequency of 0.49 in the released fish. All released fish (those with the chemical oxytetracycline tag) could be clearly and unambiguously assigned to either the genetically-tagged group of *GPI-1* 30/30* homozygotes or the non-genetically-tagged group.

Table 1. ***GPI-1**** **allele frequencies in populations of cod sampled from Masfjorden during an experiment to test the value of the rare** ***GPI-1* 30*** **allele as a marker for hatchery-produced fish**

Fish sampled	*GPI-1* 30*	*GPI-1* 70*	*GPI-1* 100*	*GPI-1* 150*
4-year-old wild cod	0.03	0.00	0.71	0.26
3-year-old wild cod	0.03	0.00	0.71	0.26
2-year-old wild cod	0.03	0.03	0.66	0.28
1-year-old wild cod	0.05	0.01	0.75	0.19
1-year-old released cod	0.49	0.02	0.43	0.06

Because equal numbers of wild type and genetically-tagged fish were released, the presence of the *GPI-1* 30* allele at a frequency close to 50% indicates that there has been no significant selection against the genetically-tagged fish—an important fact which needs to be established before such tags can be used commercially.

Identification of Diseases

One of the major constraints on aquaculture is disease. In the wild there is a natural balance between organisms and their diseases, but this balance is badly upset when organisms are cultured at high densities in enclosed situations. Entire broodstocks or complete production batches can be lost if the early stages of a disease cannot be identified and preventative measures employed. Molecular biology techniques can be used to assist in the early identification of disease-causing organisms such as viruses, bacteria and parasites, as in the following examples:

1. The mycobacteriosis disease of the European sea bass *(Dicentrarchus labrax)*, an important Mediterranean culture species, debilitates the fish, slows their growth and is

eventually fatal. Unfortunately, infected fish cannot be identified visually during the early stages of the disease. Following direct sequencing of the host and pathogen 16S rRNA genes, the disease-causing organism was identified as *Mycobacterium marinum.* Primer pairs for the 16SrRNA gene were then used in PCR trials to identify length products specific to the sea bass and products specific to *M. marinum.* A diagnostic 220 bp PCR product from *M. marinum* was identified from a particular primer pair and this method was then developed to reliably amplify the product from blood samples. This technique therefore provides a rapid non-destructive means of monitoring the broodstock of sea bass for the development of mycobacteriosis.

2. A serious viral disease of penaeid shrimp, white spot syndrome virus (WSSV), was first identified in Asia and has recently spread worldwide. The symptoms are white spots inside the carapace and a reddish body discoloration. The disease is very virulent and can wipe out a complete hatchery stock of shrimp within a week or two. Viral DNA has been sequenced and primers have been designed to amplify a short (341 bp) PCR product diagnostic for the disease from shrimp haemolymph. Not only can the presence of the viral DNA be identified, but using quantitative competitive PCR the numbers of viral genomes in an infected shrimp can be estimated. This method uses an internal standard which is co-amplified with the target viral DNA. It is amplified by the same primers but is about 50 bp shorter than the viral PCR product and the amount of this internal standard added to the PCR is known. The two PCR products are separated on agarose gel and the intensity of the stain coloration for each band is measured by a densitometer, allowing a direct estimation of viral numbers based on stain intensity.

Although not providing a cure for the disease, these types of molecular biological tool are valuable additions to the armoury that pathologists need to detect, identify and investigate the progression of diseases in aquaculture organisms.

Artificial Selection in the Hatchery

So far we have concentrated on markers at the DNA or protein level, how they can be identified and how they can be used to assess genetic variation across the genome. Now we need to consider phenotypic characteristics that are influenced by variation at one or more loci. Examples of such characteristics, or traits, would be colour variants, variations in fin size or shape, body weight at age, disease resistance, numbers of fin rays, age at maturity, larval growth rate and percentage of fat in the meat. There are two main groups of trait—qualitative and quantitative. Qualitative traits can be defined by simple discrete categories (e.g. colour variants) and are often under the control of just one or two loci, while quantitative traits require measurement or enumeration and are usually controlled by many genes (e.g. weight at age, numbers of fin rays).

Qualitative Traits

When a breeder discovers an attractive new variant, the first thing to establish is whether the variant is inherited at a single locus in a simple Mendelian way. However, there is a complication which often arises in the expression of a qualitative phenotypic trait owing to the dominant or recessive nature of the alleles controlling the trait. Let take a simple example, that of the gold colour variant in carp *(Cyprinus carpio)*, which is controlled at a single locus by two alleles, a wild-type allele (W) and a variant gold-producing allele (G). The gold phenotype is only expressed in the fish when the G allele is present in the homozygous condition (i.e. the genotype is GG), while the other two genotypes (WW and WG) will always have wild-type coloration. Conversely, the G allele is described as recessive because its phenotype is *not* expressed in the heterozygous condition.

Assuming the gold variant is of high market value, how can the breeder produce them in large numbers? Ideally need to mate a GG male with a GG female to ensure GG offspring; however, in a wild population, there are likely to be only occasional individuals expressing the gold coloration. So, to begin with, what happens if mate novel gold-coloured fish to a wild-type fish? Usually no gold offspring would appear in the F1 generation because the G allele would be rare in the natural population and a wild-type fish taken at random is unlikely to have the G allele. Without the G allele being present in both the father and the mother, no GG genotypes—and therefore no gold phenotypes—will be present among the F1 offspring. However, the F1 offspring will all be heterozygotes with one copy of the G allele and if these are mated together then this will produce GG genotypes (gold phenotypes) among the F2 generation. Assessing the ratio of gold-coloured to wild-type fish in this F2 generation should allow to identify the precise relationship between the G allele and the gold phenotype, and GG males and females from this F2 generation can be mated to produce all-gold offspring.

Although this example of gold coloration in carp is straightforward, in the real world of aquaculture things are not usually this simple. Perhaps the colour variant is sex-linked, so that it only occurs in one of the sexes. More commonly, dominance may be incomplete, such that heterozygotes will express a different colour to those expressed in the two homozygote classes. An example of partial dominance is found with the 'gold' phenotype in the tilapia species *Oreochromis mossambicus,* where the wild-type allele (W) is only partially dominant over the gold-type allele (G)—GG genotypes are gold, GW heterozygotes have a 'bronze' skin colour and WW homozy-gotes express the normal black coloration.

In some situations the two alleles at a locus are co-dominant—that is, both contribute equally to the phenotypic character of the heterozygote. This inheritance pattern is difficult to distinguish from the partial dominance of a single allele, but fortunately the consequences, from the point of view of the breeder, are the same. Obviously, to produce 100% heterozygote phenotype offspring, matings must occur between two homozygotes.

A further level of complexity occurs where two important commercial characters are controlled by two independent unlinked loci. At the first locus, the dominant wild-type allele (W) produces grey body colouration, while the recessive (G) allele produces gold coloration in homozygous (GG) fish. Because of complete dominance of the W allele, heterozygous (WG) fish are grey. At the second locus a recessive variant (C) causes curvature of the spine in CC homozygotes. The dominant wild-type allele (Z) gives homozygous (ZZ) and heterozygous (ZC) fish a normal spine.

Gametes	W Z	W C	G Z	G C
W Z	WW ZZ grey normal	WW ZC grey normal	WG ZZ grey normal	WG ZC grey normal
W C	WW ZC grey normal	WW CC grey curved	WG ZC grey normal	WG CC grey curved
G Z	WG ZZ grey normal	WG ZC grey normal	GG ZZ gold normal	GG ZC gold normal
G C	WG ZC grey normal	WG CC grey curved	GG ZC gold normal	GG CC gold curved

Ratio of different phenotypes:
9 grey normal : 3 grey curved : 3 gold normal : 1 gold curved

Fig. 1. Punnett square design showing the consequences of a mating between F1 guppies which are heterozygous at two loci (WG ZC) where the first locus controls gold colour by a recessive allele (G, wild-type =W) and the second locus controls curvature of the spine by a recessive allele (C, wild-type = Z). Phenotypes: grey with normal spine (grey normal), gold with normal spine (gold normal), grey with curved spine (grey curved) and gold with curved spine (gold curved).

The conventional way to calculate the proportions of offspring genotypes and phenotypes from matings between individuals is to use the Punnett square design where the gamete types are arrayed as column and row headers and cells within the square are simply completed by the addition of the column and row headings. In the case of guppies, following the strategy of mating between double heterozygotes, the Punnet square illustrates that the offspring phenotypes would be in the ratio of 9 wild type (grey with a normal spine): 3 gold with a normal spine: 3 grey with a curved spine: 1 gold with a curved spine.

In this example of inheritance of two characters in the guppy, there is no interaction between the two loci. The colour of the fish is not affected by the spinal curvature, and *vice versa.* Where there is interaction between loci in the expression of the phenotypes they control, this is called epistatic interaction. An example of this is the 'pearl' character of Nile tilapia *(O. niloticus)* where the phenotypic opalescent-white character of the scales (designated 'pearl') is controlled epistatically at two loci. Each locus has a wild-type allele and an alternative pearl-type allele and the pearl pheno-type is expressed only in individuals that carry pearl-type alleles from both loci. This does not involve the standard type of dominance because pearl-type alleles *can* be expressed in the heterozygous condition. It is simply the presence of the pearl-type alleles at *both* loci that confers the pearl phenotype on its carrier. This is illustrated in a Punnett square where the wild-type alleles at the two loci are W and Z and the pearl-type alleles are designated P and L respectively.

A further example of epistatic interaction involves the pattern of the scales in the common carp (C. *carpio),* one of the world's most important cultured fish. Carp with a reduced scale pattern are highly valued in the European market while Asian consumers prefer the wild-type scale pattern. At one locus (wild-type allele = W, variant allele = S) the S allele reduces the overall scaliness of the fish such that SS homozy-gotes exhibit a phenotype called 'mirror'. At the other locus (wild-type allele = Z, variant allele = N) the presence of the Z allele modifies the pattern of scales in het-erozygotes, but is lethal in the homozygous state (ZZ). This locus is epistatic to the W/S locus—it also modifies scale pattern—and it shows incomplete dominance as although ZZ homozygotes die, heterozygotes (ZN) survive and NN homozygotes have wild-type scale pattern. Figure 2 illustrates the ratios of the different phenotypes produced by mating carp which are double heterozygotes at these loci. The phe-notypes are 'normal scaled'; 'line', where the scales are mostly restricted to the lateral line; 'mirror', where a few scales are scattered across the body; and 'leather', where scales are virtually absent.

However, allelic variation at a single locus may have effects on more than one biochemical pathway and this can have subtle effects on other phenotypic characters. This is called pleiotropy and there are a number of performance-based pleiotropic effects on mirror, line and leather carp. The wild-type scaled carp generally grow, survive and resist disease better than do mirror, line or leather carp.

Gametes	W Z	W N	S Z	S N
W Z	WW ZZ lethal	WW ZN line	WS ZZ lethal	WS ZN line
W N	WW ZN line	WW NN wild-type	WS ZN line	WS NN wild-type
S Z	WS ZZ lethal	WS ZN line	SS ZZ lethal	SS NZ leather
S N	WS ZN line	WS NN wild-type	SS ZN leather	SS NN mirror

Phenotypes ratio : 6 line : 4 lethal : 3 wild-type : 2 leather : 1 mirror

Fig. 2 Punnett square illustrating inheritance of wild-type, mirror, line and leather pheno-types of carp (*C. carpio*). The mating is between fish which are double heterozygotes at two loci. One locus (wild-type allele = W, variant allele = S) controls scale density and the other, an epistatic locus (wild-type allele = Z, variant allele = N), controls the pattern of scales. Allele Z is a partially dominant lethal allele.

Globally, the development of colourful phenotypes of fish is of great economic importance. The Taiwan Fisheries Research

Institute has spent many years developing the potential of red variants of *O. mossambicus,* that are controlled by recessive alleles. They developed hybrids with *O. niloticus* and, following several generations of artificial selection, produced stable red and white strains of the hybrids. There are as yet few examples of the commercial use of qualitative traits in shellfish, such as the golden coloured Pacific oyster *(Crassostrea gigas)* from Tasmania, or the 'notata' form of the hard-shell clam *(Mercenaria mercenaria)* in the USA.

Quantitative Traits

The study of these polygenic, or quantitative, traits is the basis of quantitative genetics. Unlike qualitative trait loci, genes that are assumed to be influencing or controlling a quantitative trait are very rarely identified. They are given the overarching title of quantitative trait loci (QTL). Of course, although a quantitative trait will be influenced through its QTL it is also very likely that the trait will be influenced by the environment in which the organisms exist. A well-fed prawn will grow faster than a poorly-fed prawn, regardless of its genotype. It will also include the internal cellular environment in which the proteins produced by the genes have to operate. So when considering the influence of the environment on a particular trait it is not enough to consider only the tanks in which fish might be swimming around or simply to count the number of mysids being fed to a growing lobster larva. Environmental interactions are extremely complex and many aspects of such interactions will be invisible, undetectable, unexpected or simply unknown.

Biologists study quantitative trait variation for a number of reasons, but shall concentrate here on the use of quantitative genetics as a tool for understanding and implementing selective breeding of aquatic organisms in commercial culture. Selective breeding of some form or other has been undertaken for hundreds, sometimes thousands, of years in agricultural and domestic plants and animals. The results are all around and take them for granted. Cattle have been bred either for meat or for dairy use; sheep have been bred for wool or meat and for mountain or lowland. There are hundreds of varieties of cereals and crops that have been selectively bred for particular soils, seasons or production characteristics. Perhaps the most striking range of phenotypic variation obtained by artificial selection within a single species is evident in the domestic dog *(Canis domesticus)* ranging from the tiny Chihuahua to the giant Irish Wolfhound.

Among aquatic organisms, probably the first to be intensively bred and managed were the carps *(Cyprinus* spp.) and in the Far East intensive selection has taken place over hundreds of years for unusual colour variants leading to a highly valuable ornamental fish market. However, very few aquatic species brought into commercial culture have been subjected to the application of scientifically-based artificial selection, particularly quantitative genetics. Indeed, it is only really within the past 25 years that any studies of quantitative traits in aquatic species have been carried out, and these have been limited to a few species of fish and shellfish.

Variance of a Trait

In a sample of organisms, measurements of a continuously variable trait such as weight at age will tend to be distributed normally and the distribution can be described using standard statistical notation. There will be a mean (symbol = –), a standard deviation from the mean (symbol = s) and a variance (symbol = s^2) which is the square of the standard deviation from the mean. The variance is simply a number which represents the spread of measurements to either side of the mean: the greater the spread of measurements, the larger the variance. Some part of that variance will be the result of genetic differences between individuals and the remaining part will be the result of differences in the environments experienced by different individuals. To know the relative influences of these two factors—there is no selective advantage in breeding from the largest members of a broodstock if they are only the largest because of the environment in which they grew up.

The total variance of a trait can be divided into that produced by genes and that produced by the environment such that:

$$V_p = V_g + V_e + (Vg \times V_e)$$

where the total phenotypic variance is V_p, the variance produced by genetic differences among individuals is V_g and the component of variance produced by environmental differences is V_e. Some part of the variance will be produced by gene-environment interaction, so the term (V_g x V_e) is included in the formula. Because this component is very difficult to identify, and is thought to be usually quite small, for all practical purposes it is assumed to be negligible. There is one more element of the variance that should be mentioned. It is the variance that is the result of maternal factors, both genetic and environmental (e.g. mitochondrial DNA, egg size, brooding behaviour, nesting location), and that can represent a considerable fraction of V_g and V_e.

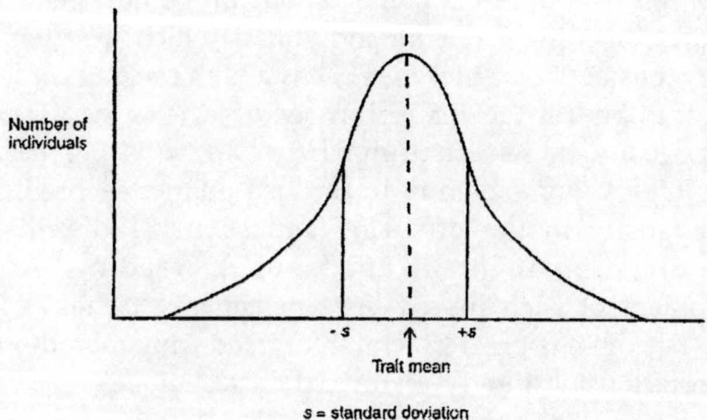

Fig. 3. Normal distribution of a trait in a sample of organisms, showing the mean and standard deviation (*s*). The variance of the trait is equal to the square of the standard deviation (s^2).

The proportion of the variance of a trait which is under genetic control is termed the heritability, denoted by the symbol h^2. The broad-sense heritability is:

$$h^2 = V_g / V_p$$

which represents the proportion of the total variance (V_p) which is produced by genetic variance (Vg).

The variation produced by genes may come from three different sources. First, the variation may be related to the straightforward presence or absence of particular alleles at the QTL. This is the additive genetic variance (V_a) and is the most important element of genetic variation for the aquaculture breeder as the presence or absence of particular alleles is a character which is passed unchanged to the next generation.

A second cause of genetic variation is due to the presence or absence of particular genotypes at QTL. For example, a particular heterozygous combination of alleles at a locus may confer an advantage or disadvantage on an individual with respect to a particular trait. This is dominance genetic variance (V_d) and is much less amenable to simple artificial selection because genotypes are broken down during meiosis and put together again in different combinations in the next generation. Therefore, breeding from an individual which, because it is heterozygous at one or more QTL, has a high ranking for a trait, does not guarantee inheritance of the advantageous heterozygosity in its offspring. However, crossing between inbred, highly homozygous lines does guarantee predictable heterozygosity in the offspring and such 'F1 hybrids' are commonly used in plant and animal breeding. As yet, development of such inbred lines in aquatic species is in its infancy but there is great potential for producing inbred lines in such species using ploidy manipulation techniques.

The third component of genetic variation is produced by interactions between loci and is called epistatic or non-allelic interaction genetic variance (V_i). Here considering that an individual may be highly ranked for a trait as it possesses particular combinations of genotypes across two or more QTL. For example, it may be an AC heterozygote at one locus, an EE homozygote at a second and an PQ heterozy-gote at a third, and it is this particular combination that gives it the high ranking.

So the genetic component of variance can be partitioned such that:

$$Vg = V_a + V_d + V_i$$

and the ratio of additive genetic variance (V_a) to the total phenotypic variance is called the narrow-sense heritability:

$$h^2 = V_a / V_p$$

Values of h^2 (either broad-sense or narrow-sense) theoretically vary from zero, where phenotypic variance is entirely the result of environmental effects, to 1.0, where all the variance is the result of genetic effects. The higher the heritability, the greater will be the resemblance between parents and offspring. It is this resemblance between related individuals that is used experimentally to estimate heritabilities of important traits.

Narrow-sense Heritability

A number of methods can be used to estimate h^2, such as regression of full-sib means on the mid-parent mean or comparison of full-sibs and half-sibs using analysis of variance. Given that much of quantitative genetics was developed around agricultural animals, the reader will find that in the literature male parents are called sires and female parents are called dams. Self-fertilisation is included in some of the experimental designs that have been used to estimate heritabilities in plants (in aqua-culture these methods could be applied to simultaneous hermaphrodites such as scallops, or, using cryopreservation of gametes, to sequential hermaphrodites such as oysters). Homozygous inbred lines are employed in other methods for the estimation of heritability and, as indicated elsewhere, development of such inbred lines is one aspect of the domestication of aquatic organisms which is likely to take place in the near future.

In order to produce reliable estimates of h^2 using analysis of variance of full-sib and half-sib hatchery crosses, all crosses must be performed at the same time and all offspring from all families must be reared in the same environment. Achieving simultaneous crosses is relatively simple in many fish and molluscs, but can be a real problem in other species. Some bivalves, such as scallops, are simultaneous hermaphrodites and great care is needed to avoid self-fertilisation in hatchery conditions. Female and male flat oysters are indistinguishable by appearance and the females brood the larvae in their mantle cavity. Many pairs of oysters therefore need to be set up to ensure that a sufficient numbers of male and female pairs. Also, the fact that individual oysters will change sex during their life is not helpful! Strip spawning of catfish males is not usually possible, so the males have to be allowed to perform multiple matings separated by varying periods of time. This results in uncontrollable differences in size or age between half-sib families and may create error in the calculation of h^2. Females of some prawn species cannot normally be induced to ovulate at a specified time and males cannot be strip spawned. In fact they deposit their spermatophore on the cephalothorax of the female just following her prespawning molt which makes control of timing of crosses very difficult. In some crustaceans, such as the freshwater crayfish, the females mate at intervals with several different males and retain their attached spermatophores until they ovulate. Consequently, not only is it very difficult to organise the timing of the crosses, but also the identity of the male parent needs to be established.

Once families have been produced, they must be kept in replicated containers in the same environment. Environmental elements such as water quality, food quality and quantity, temperature, stocking density and level of disturbance must be monitored and held constant. The potential for variation of any of these factors generally means that statistical analysis of large numbers of families is required to provide meaningful h^2 estimates. This requires a considerable investment in space and facilities which is often not feasible in a small-scale hatchery.

On the other hand, if individuals could be tagged or individually identified in some way then they could all be reared together in a common environment which removes the need for extensive hatchery space and greatly reduces the statistical problems caused by variable environments. But how can you tag individual fish or shellfish at a small size? When fish reach a certain size they can be heat- or cold-branded or their fins can be clipped, but, of course, trying to mark the carapace of crustaceans for identification purposes is useless as they moult regularly. One method used for identifying individual fish or crustacea is that of inserting a passive integrated transponder (PIT) tag into the flesh, but this would be very expensive for marking large numbers of juveniles in a commercial situation, and is really only effective for marking adult broodstock.

The development of highly variable genetic markers that can be detected from minute quantities of tissue provides a neat solution to this problem. A suite of just a few highly variable microsatellite loci is usually sufficient to enable more than 95% of individuals to be identified to their family in a 10 x 10 cross (100 families). At a highly variable microsatellite locus individual parents are likely to differ in at least one of their alleles from all other parents. When this is extended to several highly variable loci the multi-locus genotypic signal of each parent is almost certain to be distinct, and therefore the multi-locus genotypic signature of the offspring of crosses between these parents can be identified.

Although this technology is currently beyond the scope of all but the largest hatcheries there are now a number of genetic technology companies which can provide hatcheries with precise identification of their young fish or shellfish at reasonable cost. Over the coming decades further refinement of genetic methods will tend to lower rather than raise the cost of this type of activity and possibly bring it within the range of many commercial hatcheries. Apart from the observation that none of the heritabilities are greater than 0.5, and most are much less than this, two factors stand out in these figures. First, heritabilities can be very different for the different traits. In Atlantic salmon, approximately 35% of the variation in growth rate is controlled by additive genetic variation (h^2 = 0.35), but none of the variation in overall survival is so controlled (h^2 = 0.0). Secondly, even though these two species belong to the same genus, there are big differences between them in the heritability of certain traits. For example, the age at which Atlantic salmon mature appears to be more strongly under the control of additive genes than the age at which rainbow trout mature. In contrast, flesh colour has a much higher heritability in rainbow trout than in Atlantic salmon.

These observations reinforce the message that heritabilities are not easily predictable: indeed, differences in heritabilities between different populations within a species are commonly observed. Also, as might be expected, heritability estimates for a single population or broodstock will vary if trials are conducted in different environments.Variation in heritability means that some traits are far more amenable to rapid change by selection than others—the relatively high h^2 of salmon body weight at market size ($h^2 = 0.35$) implies that this trait will respond well to selection. This has been confirmed by the genetic gains in growth rate ranging from 11 to 15% per generation achieved in the National Norwegian Breeding Programme for salmon which was initiated in the 1970s.

Correlated Traits

The phenotypes of organisms comprise thousands of characters and many of these characters are interlinked in one way or another. Therefore, there is a danger that if you select for one particular trait which is of value in aquaculture, you might inadvertently be selecting for another trait which could be deleterious. For example, fast growth rate in salmon could be correlated with age at maturity such that, beyond a certain point, the advantages of fast growth will be outweighed by the tendency of these fish to mature earlier than wanted by the fish farmer. Fortunately, where such tests have been carried out in fish, high correlations have been found between feed conversion efficiency and growth rate and low correlations between growth rate, and survival. One unwanted correlation, however, is between body weight and fat content in salmonids which means that it is not easy to artificially select for increased growth rate without also increasing the body fat content.

In channel catfish *(Ictalurus punctatus)* the time of maturation is dependent on age rather than size. This can create difficulties as selected fast growing fish in some strains reach such a large size before becoming mature that they are too big to be accommodated in conventional spawning refuge chambers. However, one useful genetic correlation in this species is that between the traits of body density and dress-out percentage (the proportion of the fish's weight remaining after gutting). Normally, dressout percentage can only be assessed after slaughter but, because of the strong genetic correlation of this trait with body density, measurements of body density on living fish can be used to estimate dressout percentage.

Artificial Selection

Mass selection—also confusingly called individual selection —is the simplest type of artificial selection and involves selecting the best-performing individuals for the trait in question from a population, and breeding from them. It usually results in an increase in the mean of the trait in the offspring compared with the parental population. Mass selection can be performed for several generations until the desired shift in the mean value of the trait has been obtained. The selection differential, *s*, is the difference between the mean value for the trait in the parental population and the mean value of the individuals used for breeding. If the heritability of the trait is known, then the expected improvement in the mean value of the trait (the response to selection, *r*) can be estimated as the product of the heritability and the selection differential. In addition, if an estimated heritability is not known, a realised' heritability can be calculated following an artificial selection trial where the selection differential is known and the response to selection can be measured in the offspring.

One of the difficulties associated with predicting the response to selection is that h^2 will usually have been estimated under experimental conditions that may have been very different from the production-scale environment under which the selection is carried out. Because heritabilities may vary considerably between environments the predicted response may not be achieved. In addition, the h^2 value estimated for one population may not apply to another because the amounts of additive genetic variance for the trait may be very different between the two. Mass selection is simple to carry out and easy to manage in a hatchery situation, but it is only really effective when the true value of h^2 is greater than around 0.3 and when large selection differentials can be employed. Large selection differentials cannot be repeatedly employed when there are only small broodstock populations available because only a few individuals would be mated and inbreeding would result.

When the h^2 value for a trait is low (<0.30), improved selection response can be achieved using family selection. This involves choosing entire families, usually groups of full-sibs or half-sibs, rather than individuals for breeding. The families are selected on the basis of the mean trait value across the whole family. An individual's trait value is not considered in the selection process: the individual just has to be from a particular family. An important limitation of this method is the need to maintain a number of different families separately as broodstocks. Strategies to reduce inbreeding will also be required. Nevertheless, when h^2 is low, family selection produces a greater response than mass selection.

An alternative approach that can be usefully employed is within family selection where individuals are selected for breeding on the basis that they are highly ranked for the trait in question within their family. In this situation, the family mean for the trait is not considered. Although individuals from poorly-performing families may have a low value for the trait relative to the population as a whole, their inclusion in the breeding scheme provides an effective counterbalance to the inbreeding that can occur when family selection alone is employed. Because of this, fewer families need to be maintained and this is a real advantage in most aquaculture enterprises where the cost of holding broodstock families separately can be prohibitive. A more sophisticated method of selecting individuals for breeding is the use of marker assisted selection (MAS). During the process of gene-mapping the genetic markers used can be tested to ascertain whether possession of particular alleles at these loci is correlated with performance. Any markers that are found to show such an association (QTL) can be of value in artificial selection. MAS simply involves scoring broodstock individuals and selecting them on the basis of their genotypes at QTL.

In any breeding programme, individuals are going to be of different value depending on how much improvement in a trait their offspring produce. This is an individual's breeding value.

More than one trait of importance to aquaculture production can be used to estimate the breeding value of an individual. An individual can therefore be given an index that is the weighted sum of its breeding value for all of the traits where performance of its offspring is known. A further refinement is to weight these breeding values on the basis of the relative importance of the traits from the aquaculture point of view. Thus selection based on a carefully weighted index of the breeding value of individuals is a desirable endpoint in any breeding programme. However, considerable investment has to be made before such index-based selection is possible and it is unlikely that many aquaculture enterprises will be able to reach that endpoint. In particular, the short lives of individual fish or shellfish represent a practical limitation to development of this level of sophistication, although this could be offset by the cryopreservation of gametes. Multiple traits can be dealt with in other ways besides using breeding values. Tandem selection involves selecting for one trait in one generation and another trait in the next. Independent culling requires that each individual selected for breeding has to score well for more than one trait. Finally, if two traits are known to be highly genetically correlated then it is possible to select for one trait (the secondary trait) as a means of improving the other (the primary trait); such indirect selection is particularly useful when heritability for the primary trait is low but that for the secondary trait is high.

Realised Heritabilities

The realised heritability is only true for the particular environment in which the selection was carried out and for the particular strain or broodstock that has been selected. There are many instances where researchers in different laboratories have published very different realised heritabilities for the same species in apparently similar environments.

Breeding Programme

The initial requirement for any breeding programme is to acquire certain basic information about the population of fish or shellfish involved in the programme. First, of course, there must be a recognised method of culture that closes the life cycle. Next, there should be variation associated with the traits of commercial interest, and the extent of the phenotypic and genetic variation of these traits must be quantified. This is done by establishing, for each important trait, its range, mean, variance and heritability. When this background knowledge is in place then the breeding goals can be more precisely defined. In principle, all economically valuable traits that can be clearly measured should be included in the breeding goal. Nevertheless, breeding goals must be realistic, practicable and take account of costs, space available, generation time and predicted gain. They must be maintained in good health and their reproductive activity must be properly managed. The types of crosses to be carried out need to be carefully considered based on the type of selection to be employed (mass, family or within family) and the genetic integrity of all families of offspring needs to be strictly maintained within the hatchery and under nursery culture.

There is always the possibility that artificial selection produces little or no gain. This is often the case and can be the result of there being limited genetic variation within the broodstock or the population being selected. Even if h^2 is high, in the absence of variation to exploit there can be little progress from generation to generation. One explanation for the lack of variation could be that the population was the result of an initial founder event and that many subsequent generations of significant inbreeding have resulted in high homozygosity across the genome. One obvious solution is to introduce new genetic variation into a broodstock by introducing individuals from alternative populations or broodstocks. However, care needs to be exercised to ensure that this does not risk the unintended introduction of non-native individuals into the wild.

Inbreeding, Cross-breeding and Hybridisation

Dominance effects represent interactions between pairs of alleles at the same locus but genotypes are disassembled at meiosis and reassembled randomly during syn-gamy. Because of this, when breeding from a population cannot guarantee the exclusive production of particular genotypes except in certain circumstances. For instance, breeding from a broodstock of animals that are all homozygous for a single fixed allele at a locus will produce all-homozygous offspring. But if some of the broodstock are heterozygotes then it is not possible to produce all-heterozygous or all-homozygous offspring.

These qualities are therefore directly related to dominance genetic effects. The expression of heterosis, where highly heterozygous offspring exhibit increased fitness, is obviously an important goal of hatchery production. One of the simplest ways to do this is to develop different strains or lines of highly homozygous individuals for crossing. But, of course, such highly homozygous lines risk inbreeding depression and may be difficult to grow and maintain in the hatchery. An alternative method is to cross between different natural strains, or even between closely related species. All offspring from such crosses or hybridisations will have identical heterozygous genotypes at loci that are homozygous for different fixed alleles in the parents.

Evidence for the superior fitness of cross-bred or hybrid offspring is common in agricultural crops and animals. In many species and for many traits they significantly exceed the best-performing parental line. Nevertheless, there are also plenty of examples where the crossed phenotypes are only intermediate between, or even inferior to, the two parental phenotypes. Hybridisation and crossing between strains are therefore not guaranteed to be successful in improving the performance of offspring. Nevertheless, where heterosis has been produced in cross-bred offspring some of this heterosis will remain in subsequent generations as long as random mating occurs and the effective population size is sufficiently large to avoid inbreeding.

Where a trait has a very low narrow-sense heritability (low V_a), cross-breeding can be used as an alternative to selection for that trait. However, because dominance effects are not predictable, the potential performance of an individual strain used for cross-breeding can only be assessed on the results of a number of line-by-line crosses. In this way the mean performance of the cross-bred offspring of a particular line in relation to the average performance of cross-bred offspring from all possible lines can be assessed. This is the general combining ability of the line, while its specific combining ability is a measure of the performance of its offspring when crossed with an individual line. Currently, developments in the domestication of most aquaculture species have not reached the point at which sufficient numbers of inbred lines or strains are available to consider their combining abilities.

Hybridisation between closely related species can be seen as simply an extension of cross-breeding between strains or lines within a species. Although, strictly speaking, appropriate only for interspecies crosses, the terms 'hybridisation' and 'hybrids' are nevertheless often used rather loosely about cross-breeding within species. Hybridisation trials have been carried out in a number of groups of aquaculture organisms with varying success. When the species differ in their mechanisms of sex determination, the hybrids will often be of only one sex. The production of monosex hybrids can be a very valuable feature because it enables better control of grow-out as animals reach maturity. Where both sexes are present during grow-out, often a considerable amount of the energy provided in the feed goes towards the costs of fighting, courting and copulation. In some cases hybrids are effectively sterile and this may be advantageous.

It is also important to consider species' differences in maternal effects such as mitochondrial DNA, egg quality and brooding. For example, in hybridisations between the striped bass *(Morone saxatilis)* and the white bass *(Morone chrysops)* best results were obtained by using females of the larger striped bass crossed with male white bass. This is because the striped bass produce more eggs and the hybrid larvae from this direction of cross are larger and more hardy. However, sometimes in the real hatchery factors other than genetics control what can be achieved—adult female striped bass are not all that easy to collect nor to maintain in good spawning condition so the industry has had to compromise and use the less productive direction of cross, white bass females with striped bass males.

In oysters of the genus *Crassostrea,* hybrid crosses will usually produce larvae but in most cases these do not survive. Only the cross between *C. rivularis* and the Pacific oyster, *C. gigas,* seems to produce viable offspring, and even then there is a difference between the directions of hybridisation. Male *C. gigas* crossed with female *C. rivularis* produces feebler offspring than *C. gigas* eggs fertilised with *C. rivularis* sperm. In no cases so far have oyster hybrid offspring grown faster or survived better than either of the pure species lines in properly controlled experiments, so this does not look like a particularly rewarding avenue of further research effort. In spite of this, there is still the potential that hybridisation could bring together other qualities from the parental species, for example salinity tolerance or disease resistance, which could be more highly valued than simply growth performance.

Within the salmonids as a group, many attempts have been made to improve performance by hybridisation or cross-breeding between strains, but with mixed results. Crossing between two strains of *Oncorhynchus mykiss,* the freshwater (and effectively domesticated) rainbow trout and the anadromous steelhead trout, produced a cross-bred strain which had reduced seasonal variation in seawater adaptability compared with the pure steelhead trout. Similarly, stocking success of *Salmo trutta,* the brown trout, has been improved by cross-breeding between a domestic and a wild strain, while crosses between certain domestic strains can produce hybrids with improved growth rates. However, hybridisation trials between arctic charr *(Salvelinus alpinus)* and brook charr *(Salvelinus fontinalis)* demonstrated that hybrids were viable, but that there was no evidence for heterosis in growth or any other commercial trait.

In spite of the mixed results seen in salmonid interstrain crosses and hybridisations, significant and therefore commercially valuable heterosis has been demonstrated in a number of other fish species, such as sturgeons *(Acipenser ruthenus* x *Acipenser baeri),* tilapias (Nile x red; Stirling Nile x

Korean Nile; Stirling Nile x Japanese Nile) and African catfish *(Clarias anguillaris* x *Clarius gariepinus)*. These mixed results of hybridisation trials illustrate that although there is the potential for genetic improvement through hybridisation, success is very hard to predict and considerable research needs to be done on a case by case basis before commercial application is considered. Nevertheless, one prediction that can be made is that successful hybridisation between species that have different chromosome numbers, or strongly differing karyotypes, is very unlikely. Therefore a detailed knowledge of the karyotypes of potentially hybridising species is of importance for this type of development in aqua-culture.

While originally only valued as ornamental fish, tilapias, or 'aquatic chickens' as they are sometimes known, are now an extremely important international aquaculture product, especially in Africa (their region of origin) and in the Far East. Extending this approach, ICLARM has set up an International Network on Genetics in Aquaculture (INGA), which fosters research into the genetics of species such as the catla *(Catla catla)*, rohu *(Labeo rohita)* and silver barbs *(Barbodes gonionotus)* in Bangladesh and India, the common carp *(Cyprinus carpio)*, silver carp *(Hypopthalmichthys molitrix)* and tilapias in China, the mrigal *(Cirrhinus mrigala)* in Vietnam, the freshwater prawn *(Macrobrachium rosenbergii)* in Malaysia, and tilapias in many African and Asian countries.

References

Benzie, J.A.H. (ed.) (2002) *Genetics in Aquaculture VII*. Elsevier, Amsterdam.

Falconer, D.S. (1989) *Introduction to Quantitative Genetics*, 3rd edn. Longman, New York.

Gjedrem, T. (2000) Genetic improvement of cold-water fish species. *Aquaculture Research*, 31, 25–33.

Kearsey, M.J. & Pooni, H.S. (1996) *The Genetical Analysis of Quantitative Traits*. Chapman & Hall, London.

Lutz, G.C. (2002) *Practical Genetics for Aquaculture*. Blackwell Science, Oxford.

Tave, D. (1993) *Genetics for Fish Hatchery Managers*, 2nd edn. Van Nostrand Reinhold, New York.

5

Polyploid Induction in Fish

The polyploid state refers to individuals with extra sets of chromosomes. The normal and most common chromosome complement is two sets (diploid). Triploidy refers to individuals with three sets of chromosomes and tetraploidy refers to individuals with four sets. Hexaploids have six sets, and aneuploids have at least a diploid set with one or more additional chromosomes, but not a full complement, to the set. Polyploidy is lethal in mammals and birds, but has led to the development of many useful, improved plant varieties, including domestic wheat.

Usually gametes have a single set—they are haploid—and two gametes fuse to form a diploid zygote. However, it is viable for organisms to possess three (triploid), four (tetraploid) or more copies of each chromosome, in which case they are known as polyploids. Polyploidy can happen in nature—one example being hexaploidy in wheat—but it can also be induced. Because the eggs of most fish and shellfish are released into water before fertilisation it is relatively easy to access the maturation divisions of the egg and the early divisions of the embryo. Such access allows manipulation to create polyploid embryos and can also be used to produce diploid embryos that contain only maternal chromosomes—gynogens, or only paternal chromosomes—androgens. This is one field of biotechnological development in aquaculture that does not have its origins in agriculture, given the difficulty in accessing the egg in agricultural animals and plants.

Thus, in males, four haploid spermatozoa are produced from each diploid germ cell. However, in females, both meiosis I and meiosis II produce daughter cells of such very uneven size that the smaller cells (the polar bodies) often do not complete meiosis, therefore only one haploid egg and two (rarely three) polar bodies are formed from each diploid germ cell. It is not just that fish and shellfish eggs are released into water that makes ploidy manipulation possible, but rather that the meiotic divisions have not been completed when the eggs are spawned. In the case of most molluscs, eggs are released at metaphase of meiosis I. The first polar body has yet to be produced, and an even more convenient feature is that activation by spermatozoa is required before meiosis will proceed. In most fish the eggs are released after meiosis I with the first polar body present, but further development through meiosis II is again dependent on activation by spermatozoa.

We should note that, although there is external fertilisation in most fish and molluscs, this is not generally the case in crustacean shellfish nor in gastropods such as the whelk, and that this limits the potential of ploidy manipulation in these groups. Also, among the bivalve molluscs, the flat oysters (e.g. the European native oyster, *Ostrea edulis)* retain their eggs and brood their embryos in the mantle cavity.

The basic idea behind the method of ploidy manipulation is to allow chromosome replication, but prevent cell division. In that way daughter cells have double the number of chromosomes. If are able to suppress the cytoplasmic division of meiosis I then the resultant single cell will contain all the pairs of chromosomes and no first polar body is formed. What would normally be two haploid cells will be a single diploid cell. If, on the other hand, meiosis I is allowed to proceed as normal and cytoplasmic division in meiosis II is suppressed instead, then the chromosomes normally expelled into the second polar body are retained and a diploid cell again results. Thus, manipulating ploidy at meiosis I *or* meiosis II makes the egg diploid rather than haploid. Subsequent syngamy with the haploid male pronucleus results in a triploid embryo. In fish, meiosis II is always the division targeted for the production of triploids because meiosis I has been completed before spawning. In molluscs, while it may not initially appear to matter whether the triploids produced come from suppression of meiosis I (MI triploids) or meiosis II (MII triploids), there is a potential effect on heterozygosity, which will consider later.

Another trick is to allow both meiotic divisions to take place, and to allow the male and female pronuclei to unite in syngamy, but to suppress the first cleavage division of the zygote. This again produces a daughter cell with double the number of chromosomes—a tetraploid. Although tetraploid fish are produced using this approach, it is far less successful in molluscs; alternative methods have been developed for the tetraploidisation of oysters.

Production of Gynogens and Androgens

In gynogen production, sperm are treated for a short time with X rays, gamma rays, or, most commonly, by UV irradiation. This effectively breaks up the chromosomes but does not destroy the motility or the ability of such sterile spermatozoa, or sterizoa, to activate eggs. Eggs activated by sterizoa would normally result in haploid embryos and these are invariably nonviable. However, in molluscs, any one of the three divisions—meiosis I, meiosis II or first cleavage—can be suppressed in order to double the chromosome number, while in fish, meiosis II and first cleavage can be targeted to ensure diploidy. In all cases, the diploid embryos produced contain only the chromosomes of the female parent. Because there is no contribution from male chromosomes inbreeding is high, but the level of inbreeding in gynogens will depend on which division is targeted. Doubling the chromosome number at first cleavage results in homozygosity at every locus in the genome. That is not to say that all the embryos are identical clones, but each individual is homozygous at all of its loci and thus 100% inbred. These mitotically-pro-duced individuals are called mitogynes and, because all loci are homozygous, any deleterious recessive alleles are exposed with consequent reduction in viability. Gynogens produced by suppression of meiosis I or meiosis II are called meiogynes and they will not be entirely inbred because of recombination events occurring during meiosis.

Androgens are individuals whose chromosomes are entirely paternal. They can be produced by two methods, both of which require that eggs are irradiated to destroy their chromosomes. The first uses normal haploid sperm to fertilise the irradiated eggs and then the first cleavage division is suppressed to produce a diploid embryo. The second employs diploid sperm—the gonad product of a tetraploid male—to fertilise the irradiated egg and diploid embryos are produced without further treatment.

Identification of Ploidy Change

Production of triploids by physical or chemical suppression of divisions seldom results in 100% triploids and production of tetraploids is even less certain. For this reason, hatcheries need to establish how successful their triploid or tetraploid induction efforts have been. A balance must be struck between the proportion of triploids in a batch and the costs of ensuring triploidy induction. There are a number of ways in which the ploidy of individual fish or shellfish can be identified and the quickest method is to use a flow-cytometer, which measures the amount of DNA-specific stain taken up by nuclei of individual cells in a sample. A graphical display allows calculation of the proportion of individuals in the sample that are triploid and the proportion that are diploid (or of other ploidies) by measuring the areas under the peaks coinciding with the amount of DNA present in the nuclei of each ploidy type. The main problem with this method is that the price of a flow-cytometer is prohibitive for most hatcheries. Nowadays, however, there are companies that have specialised in the production of triploids of one or two of the main commercial aquaculture species and they will have their own flow-cytometers. Smaller producers can buy in ready-produced triploid seed or fry. In addition, research institutions and private companies with these facilities are not averse to providing a costed service for smaller-scale operators who may wish to carry out their own ploidy modification work.

Other methods of identifying triploids and tetraploids include direct chromosome counting, sizing of nuclei of cells from blood or other tissues and scoring genetic markers. Chromosome counting requires experience and can be quite time consuming, but has the advantage that it is a direct method. Using blood for nuclear sizing is cheap, rapid and effective and can be non-destructive but nuclear sizing of the cells of other tissues requires the sacrifice of the sampled animals. Use of genetic markers such as allozymes for ploidy confirmation relies on the fact that some polyploids will exhibit more than two alleles at highly polymorphic loci. For example, a triploid could express three alleles, with the genotype ABC. In theory, a triploid heterozygote with only two alleles (e.g. AAB) could be distinguished from a diploid (AB) by the stronger staining of the band produced by the double allele, but in practice this is not very reliable. However, the use of microsatellite loci is a great improvement on allozymes because (being highly heterozygous) they are more likely to show three bands and (because minute amounts of DNA are required) they can be scored by non-destructive sampling.

Confirming that diploid gynogens or androgens have been successfully produced is a bit more tricky than confirming the presence of triploids or tetraploids. Obviously, there is no difference in chromosome number, nuclear size or DNA content between normal diploids and gynogens or androgens. Here will consider gynogens, but similar methods apply to androgens. Remember that gynogens are produced by activating eggs with sterizoa and then using a shock to suppress a cell division (meiosis I, meiosis II or first cleavage). One method that provides indirect confirmation of gyno-gen production is to use two controls without the shock to suppress cell division—one using untreated spermatozoa and the other using sterizoa. If everything has worked properly (sperm irradiation and the shock method) then the untreated spermatozoa control will produce normal diploid offspring while the sterizoa control will produce haploid embryos that do not survive. If these controls show the expected results then any surviving embryos from the fully treated group should be gynogens.

Sometimes the spermatozoa of a closely related species can be used to fertilise eggs, producing non-viable hybrids. In this situation sterizoa of the closely related species can be used in gynogen production. Given that hybrids are non-viable, any surviving offspring must be gynogens.

Alternatively, genetic markers can be used to confirm individuals as gynogens. For example, albinism is a recessive trait and spermatozoa from a male trout which is homozygous for pigment production (AA) can be used with eggs from an albino (aa) trout. Gynogens will all be albino (aa) and normal diploids will be pigmented (Aa). Allozyme or microsatellite loci can also be used to identify gynogens when the allele(s) present in the spermatozoa are different to the allele(s) present in the egg.

Triploid Production

Triploid fish are viable and are usually sterile, which is a result of lack of gonadal development. Culture of triploid fish can be advantageous for several reasons. The potential of increased growth, increased carcass yield, increased survival and increased flesh quality are the main culture advantages. At the onset of sexual maturity, reduced or inhibited gonadal development may allow energy normally used in reproductive processes to be used for growth of somatic tissue. The sterility of triploids would be desirable for species such as tilapia, where excess reproduction may occur in production ponds. Use of sterile triploids can prevent the permanent establishment of exotic species in otherwise restricted geographical locations. Induction of triploidy in interspecific hybrids can prevent the backcrossing of hybrids with parental species, and also allows viability in some unviable diploid hybrids. Other potential uses include supplemental stocking of natural populations without compromising the genetic integrity of the resident population, disruption of reproduction in nuisance species and sterilization of transgenic fish, all mechanisms for reducing environmental risk and applying genetic conservation. The performance of triploid fish is dependent on the species and age of the fish as well as the experimental conditions. Genotype–environment interactions are common for triploid fish.

Triploid production has great potential to enhance performance in fish and shellfish; however, many problems exist. The first is that triploids can sometimes be fertile, defeating the advantages of sterility. Polyploidy can decrease performance for some traits. For many species, polyploid production may not be economically feasible.

Production of triploids is the most common target of ploidy manipulation attempts and they have a variety of uses in aquaculture. Triploids have been produced, at least experimentally, in almost all commercially aquacultured fish species, including carps, catfish, *Tilapia* and salmonids, and in most molluscan shellfish groups such as oysters, clams, scallops and abalones. One important reason for producing triploids is that they are sterile. During early gametogenesis, germ cells begin the process of meiosis and it could be that attempts to pair up the chromosomes in early meiosis are impeded by the fact that there are three homologous chromosomes rather than two.

Whatever the reason, gonad development is always much reduced in triploids. Some eggs and spermatozoa *can* develop in triploids, but these are seldom of normal ploidy and are usually aneuploid (missing chromosomes or parts of chromosomes). Triploid sterility means that, as triploid fish or shellfish reach maturity, energy that, in diploids, would go to developing gametes is available for somatic growth. As they get older, therefore, triploids effectively grow faster than diploids. Whether this potential for increased somatic growth is realised as a reduced time to market size depends on a number of factors, not least the species being cultured. Triploid fish and shellfish generally demonstrate a clear growth advantage over diploids during annual gametogenesis, but there may be more than one year of maturity before market size is reached and the advantage from the first year may not be carried over to the next. Similarly, superior growth performance of triploids in many fish species has only been demonstrated for a small part of their life histories.

Apart from the actual or potential increase in somatic growth, there are other advantages that accrue from the sterility of triploids. Sterility makes feasible the aquaculture of non-native species or genetically depauperate hatchery-produced stock, either of which might otherwise cause adverse environmental impact if they or their gametes were to escape into the wild. Before beginning such an enterprise it is necessary to confirm, first, that all individuals are triploid and, second, that even if they produce some gametes these will never produce offspring. Considering the first requirement, methods are needed to ensure 100% triploid production and also to assess the ploidy of every individual that is to be introduced. Rather than try to perfect other triploidy induction methods, the development of tetraploids as broodstock has been the favoured option. A tetraploid individual will produce diploid gametes and these, when combined with haploid gametes from a normal diploid, produce 100% triploids. Triploids produced by this method have been called interploid triploids. Generally tetraploid males are more valuable because the diploid spermatozoa produced can be used to fertilise the eggs from a large number of females. Even though this method is quite certain to produce only triploids, regulatory bodies are wise to insist on certification that all individuals for importation are triploid, based on flow-cytometry or some other proven method, and that any gametes produced from triploid individuals are effectively non-viable.

The licensed importation of non-native, triploid grass carp into many states in the USA was only allowed when research had established that, even after induction of gametogenesis by hormonal injection, almost all the spermatozoa released by these triploids were abnormal. Calculations indicated that the probability of triploid grass carp having fertile offspring was so low that they could be considered to be sterile for all practical fishery management purposes.

Another example of the importance of regulation and monitoring of the introduction of triploids of non-native species concerns triploid Pacific oyster *(Crassostrea gigas)* on the east coast of the USA. The backgound to this is that the American oyster *Crassostrea virginica* has suffered severe declines over recent decades, mainly caused by two protozoan parasites—dermo *(Perkinsus marinus)* and MSX *(Haplosporidium nelsoni).* Several eastern USA states had considered the possibility of introducing the Pacific oyster, which is not affected by these diseases; as an alternative for the local oyster culture industry. A trial introduction programme of triploid *C. gigas* was instigated, in which the triploid nature of every individual was confirmed by flow-cytometry of blood samples. After a year or so, as oysters were retested for ploidy confirmation, it was discovered that some oysters (or at least some tissues from some oysters) had reverted to diploidy. Obviously this was a serious blow to the plans to use triploid oysters in this way, and further research is continuing to explore this unexpected phenomenon.

In spite of uncertainties concerning reversion to diploidy, there is one area where triplod *C. gigas* have been a commercial success. Pacific oysters (*C. gigas)* were first introduced from Japan into the Pacific north-west of the USA in the middle of the last century and soon became naturalised. They have formed the basis of a significant hatchery-based aquaculture industry supplying Pacific oysters to the market at a time when production of the native American oyster (*C. virginica)* was in decline. However, there is a problem with Pacific oysters which is less significant in *C. virginica:* they become unmarketable when they reach sexually maturity in the summer. The reproductive tissues ramify throughout the oyster's body and its glycogen stores are converted into gametes. This changes the normal sweet flavour of the oysters and significantly reduces the quality of both taste and texture. In triploid oysters little gametogenesis takes piace and stored glycogen remains unconverted and they therefore provide a year-round marketable product. With the successful recent development of tetraploid Pacific oysters, commercial supplies of 100% triploid oyster seed are now readily available and most commercial production of Pacific oysters in the USA is now of triploids.

A further potential value of triploidy in bivalves concerns the production of artificial pearls from pearl oysters *(Pinctada* spp.). The method requires that a 'seed' pearl is implanted into tissue close to a vestigial loop of the gut, but it is a tricky operation and usually results in some mortality and some implant rejections. In ripe individuals, the gut is surrounded by gonadic material which adds to the difficulty of implantation, so sterile triploids could be used to reduce mortality and implant rejections.

In salmon aquaculture, triploids have been proposed as a method to overcome problems associated with grilsing. Grilse are fish (both male and female) that mature after only one year at sea, leading to a deterioration in flesh quality and a transfer of energy from somatic to gonadic growth as maturity approaches. Identifying, sorting and trying to market grilse creates practical difficulties, and the use of sterile triploid salmon was an approach considered by the industry. However, strains of salmon that have been artificially selected for late maturity are now available and these have become the preferred means of reducing grilsing in salmon farms.

The process of triploidisation can also be of value when interspecific hybrids are produced. For some reason, possibly the presence of a double set of chromosomes from one or other parent species, interspecific hybrid triploids can often outperform their diploid counterparts. Indeed, in some circumstances triploidisation of hybrids may be the only way to produce survivors of interspecific crosses.

For some shellfish, such as sea urchins and some scallops (in Europe), the roe or gonad is the important market product. In these cases there is obviously no commercial advantage to producing triploids.

	Without recombination	With recombination
Diploids	AC or BC	AC or BC
Meiosis I triploids	ABC	AAC, BBC or ABC
Meiosis II triploids	AAC or BBC	ABC

Fig. 1. Genetic consequences of triploidy induction in relation to heterozygosity. Potential genotypes of diploid and triploid offspring from a female heterozygous for alleles A and B, mated with a male homozygous for allele C.

Although the advantages that accrue to triploids through their sterility are the most important ones from the aquaculture point of view, should also consider the question of the potential effect of increased overall heterozygosity of the genome of triploids compared with diploids. Genetic theory is steeped with the notion that heterozygosity is an advantage over homozygosity, and that this is linked with the phenotypic phenomena of hybrid vigour and inbreeding depression. Accepting this, there is nevertheless a difficulty in mapping these effects back to heterozygote advantage at more than a very small proportion of single gene loci in the genome as is allowed under neutral theory.

Tetraploids

The principal value of tetraploids is that they can be used as a source of diploid gametes in the production of interploid triploids and androgens. Tetraploid fish have been successfully produced by targeting the first cleavage division, but this approach has not been productive in molluscs. It is possible that the differences in early embryonic divisions in these two groups may account for this. In fish the early cleavage divisions are equal. In contrast, the two-cell stage in bivalve molluscs consists of two unequally sized nucleated cells with a cytoplasmic 'polar lobe' extruded from the larger of them. The polar lobe is resorbed back into this cell during second cleavage. Suppressing first cleavage, if it interferes with polar lobe formation, is likely to have considerable consequences on early embryonic development and might explain the non-viability of tetraploids produced in this way.

Because of the failure of suppression of first cleavage to produce viable tetraploid molluscs, researchers have sought an alternative approach. Small numbers of tetraploid broodstock of the Pacific oyster have now been produced by suppressing the meiosis II division in (rare, often aneuploid) eggs from triploid oysters and fertilising them with ordinary haploid spermatozoa. A number of different ploidies, including a few tetraploids, are produced by this method and, unlike tetraploid embryos induced by suppression of first cleavage, some of these tetraploids survive. It is suggested that the extra cytoplasm present in a triploid egg is needed to support a tetraploid nucleus and that this explains why these tetraploids are viable.

Adult male tetraploid fish and oysters have now been used successfully to produce interploid triploids, although diploid spermatozoa are less active and less viable than normal haploid spermatozoa. Eggs from female tetraploid fish have also been successfully crossed with normal haploid spermatozoa.

Obviously, using diploid sperm from tetraploid males to fertilise diploid eggs from tetraploid females will result in all-tetraploid offspring, and tetraploid broodstock lines have been developed for rainbow trout *(Oncorhynchus mykiss)* and some other fish. It is possible that in tetraploids some of the potential dangers of the homozygous exposure of deleterious recessive genes as a result of inbreeding in an enclosed broodstock might be ameliorated by the fact that there are four copies of each chromosome. On the other hand, nothing is known about the long-term genetic, physiological or biochemical consequences of continual inbreeding of a tetraploid stock.

Gynogens and Androgens

Diploid gynogens are individuals that contain two copies of the maternally inherited chromosome set. One of the most useful features of gynogens is that in species where there is an XY sex-determining system, all gynogens will be XX and therefore all female. Monosex populations of fish can also be produced by hormonally-induced sex-reversal, but gynogen production avoids the use of chemicals and produces geno-typic, rather than 'phenotypic', females. If some of the all-female gynogens are then hormonally sex-reversed to phenotypic males, they can then be mated with other gynogens to produce all-female offspring without having to go through the process of gynogenesis again.

Simple chromosomal sex determination in fish may be by the XY system, where the female is the homogametic sex (XX), or by the WZ system, where the male is the homogametic sex (WW). However, sex determination is usually more complex than a simple chromosomal system as several autosomal gene loci are also involved. Gynogens can be used to explore whether a simple chromosomal sex-determining mechanism is present in a species: if half are male and half are female this would point to a WZ system; if all are female the XY system is indicated.

The sex-determination mechanism of molluscs is not well understood, but certainly does not consist of a simple chromosomal system. Indeed, many bivalve molluscs are hermaphrodites and some, including oysters, are sequential hermaphrodites, usually changing from male to female. So the study of gynogens is not expected to be much help in elucidating the sex determination mechanisms in bivalve molluscs.

A further important value of gynogens is that they are strongly inbred and have the potential to provide the basis for inbred lines in just a single generation. Mitogynes are expected to be 100% homozygous at all their gene loci while meiogynes will be less inbred, and the level of homozygosity at particular loci will depend on the frequency with which those loci are involved in recombination events. Such high levels of inbreeding result, as expected, in inbreeding depression. Gynogens generally have higher mortality and slower growth than normal diploids, but if they survive to maturity they can provide an inbred line for later outcrossing.

There is a further potential use of gynogens. If it is possible to produce mitogynes that reach maturity and produce eggs, and these eggs can again be treated to produce mitogynes, then these offspring will be genetically identical clones of one another. They will be homozygous for the same allele at every locus. The experimental value of such clones of a sexually reproducing species is high.

In a further twist, it has proven possible to develop 'supermale' tilapia individuals that have two Y chromosomes but no X chromosome. These YY supermales produce only male offspring and therefore provide the complement to female mono-sex production.

	Without recombination	With recombination
Meiosis I	AB	AA or BB
Meiosis II	AA or BB	AB
First cleavage	AA or BB	AA or BB

Fig. 2. The genetic consequences of gynogenesis induction on heterozygosity. Eggs are from a female heterozygous for alleles A and B.

Diploid androgens differ from gynogens in that both chromosome sets are obtained from the male parent. The uses of androgens are similar to those of gynogens. In fish with a WZ sex-determination system, androgens will all be male (WW) and this therefore enables monosex production. Alternatively, in fish with an XY sex-determination system, half of the androgenic offspring will have two Y chromosomes and, if they survive to sexual maturity, will be YY supermales.

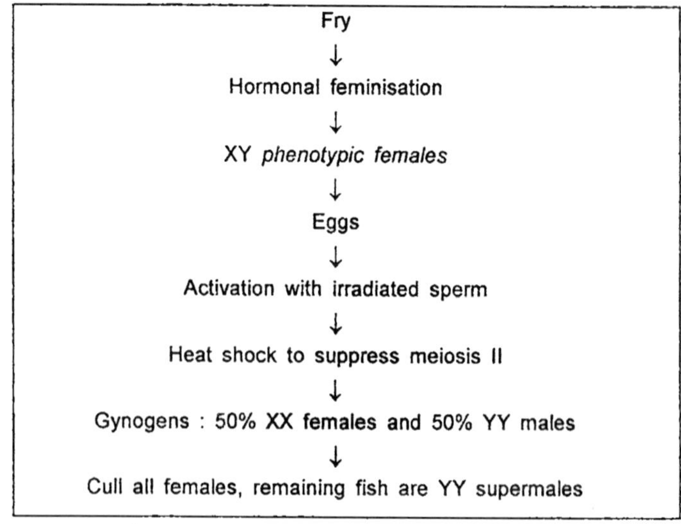

Fig. 3. Method for the production of YY supermale tilapia.

Although there are all these possibilities for the use of androgens and gynogens, not all have yet been realised. Research has been mainly directed at developing and refining methods for their production; their use in commercial hatcheries is not yet routine.

References

Beaumont, A.R. & Fairbrother, J.E. (1991) Ploidy manipulation in molluscan shellfish: a review. *Journal of Shellfish Research,* 10, 1–18.

Bidwell, C.A., Chrisman, C.L. and Libey, G.S. (1985) Polyploidy induced by heat-shock in channel catfish. *Aquaculture* 51, 25–32.

Lutz, G.C. (2002) *Practical Genetics for Aquaculture.* Blackwell Science, Oxford.

Purdom, C.E. (1973) Induced polyploidy in plaice (*Pleuronectes platessa*) and its hybrid with the flounder(*Platichythys flesus*). *Heredity* 29, 11–24.

Schultz, R.J. (1980) Role of polyploidy in the evolution if fishes. In: Lewis, W.H. (ed.) *Polyploidy: Biological Relevance.* Plenum Press, New York.

6

Genetic Engineering in Aquaculture

Genetic engineering, genetic modification or genetic manipulation—all three terms mean the same thing, the reshuffling of genes usually from one species to another; existing examples include: from fish to tomato or from human to pig. Genetic engineering (GE) comes under the broad heading of biotechnology.

Genetic engineering (GE) is used to take genes and segments of DNA from one species, e.g. fish, and put them into another species, e.g. tomato. To do so, GE provides a set of techniques to cut DNA either randomly or at a number of specific sites. Once isolated one can study the different segments of DNA, multiply them up and splice them (stick them) next to any other DNA of another cell or organism. GE makes it possible to break through the species barrier and to shuffle information between completely unrelated species; for example, to splice the anti-freeze gene from flounder into tomatoes or strawberries, an insect-killing toxin gene from bacteria into maize, cotton or rape seed, or genes from humans into pig.

DNA Construct

The initial step in transgenesis (creation of a genetically engineered organism) is to design and build a DNA construct. Genetic engineering generally has the aim of inserting DNA which enables the genetically modified organism (GMO) to produce a protein in a specific tissue at appropriate times and at a sufficient concentration. The construct therefore must consist of the DNA to code for the protein of interest (the transgene) together with DNA coding for regulatory elements that direct gene expression (the promoter).

The Transgene

The most commonly employed genes so far in fish transgenesis are growth hormone (GH) genes—the specific GH genes of humans, rats, cattle, chinook salmon, rainbow trout, tilapia and sea bream have all been used in attempts to increase growth rates and thus reduce time-to-market size. Another gene which is of special relevance to aquaculture in polar regions is the antifreeze protein gene which is found in fish such as ocean pout or winter flounder that can live in subzero temperatures. As mentioned above, if other species could be genetically modified to produce an antifreeze protein then this could extend the area available for aquaculture. Transgenesis is also being developed as a way to induce immunity to disease-causing organisms in fish. It is possible to induce immunity somatically by injecting DNA that codes for viral protein directly into muscle tissue, but further research on such genetic vaccines will be required before veterinary approval. Inserting the gene for the viral protein into the germline, however, would mean that lines of fish could be produced with built-in immunity to a particular virus.

As far as aquacultured invertebrates or plants are concerned, there does not appear to be any current commercial transgenesis. There is research interest, however, particularly on the use of genetic modification to produce disease-resistant bivalve molluscs.

When a gene is developed as part of a construct for transgenesis it is not usually taken directly from the donor's DNA. These introns are spliced out during production of mRNA. If mRNA is extracted from cells or tissues where the gene is expressed, then complementary DNA (cDNA) can be synthesised from the mRNA by the enzyme reverse transcriptase. This cDNA can then be inserted into a plasmid vector and cloned into *E. coli* in the normal way to produce a cDNA library. The cDNA library is then screened using a probe based on the DNA sequence of the gene (or even the sequence from a related species or group, if the sequence is unknown for the species in question) to identify the matching cDNA. The cDNA is then snipped out of the plasmid and spliced to the promoter sequence.

In some cases it appears that certain introns are required for the full expression of the gene. This implies that some introns may have a regulatory function even though the sequence is not translated.

The Promoter

There are estimated to be from 30 000 to 100 000 genes in the vertebrate genome but only about 10%, mainly housekeeping genes involved in simple cellular functions, are switched on and expressed in every cell all of the time. In addition, perhaps a few thousand more genes are involved in directing activities specific to a cell's particular function. So the majority of genes are not normally switched on in a cell and a critical requirement of transgenesis is that the gene inserted into the GMO be expressed. For this reason any gene sequence in a construct is accompanied by a promoter. The promoter consists of DNA sequences that direct transcription and translation of the gene into protein. The promoter sequence regulates when, where and how much expression of the gene will occur. Early experimental work on fish transgenesis used promoters from avian (Rous sarcoma virus, RSV) or primate viruses (simian virus, SV40; cytomegalovirus, CMV) but recently, efforts have been made to design an 'all-fish' promoter from fish genomes rather than the genomes of other groups.

Once the DNA construct has been formed by splicing together the gene and the promoter, it must be amplified to produce the billions or trillions of copies needed for each transgenesis attempt. The method used is to insert the construct into a plasmid which is replicated in *E. coli* by standard cloning methodology. In most experimental trials, and certainly in any commercial application, the plasmid DNA is removed before transgenesis because this prevents the insertion of prokaryote plasmid DNA into the target organism. The potential effects of prokaryote DNA on eukaryotic transcription and translation processes include methylation (replacement of hydrogen by a methyl group, CH_3, in a DNA base) but these effects are not predictable and must therefore be avoided. Of course, there are hundreds of millions of bacterial genomes normally in intimate contact with organisms in their gut, gills and exterior surfaces, but these intimate associations have evolved, and are continuing to evolve, by natural selection and can therefore be regarded as 'normal' and environmentally balanced.

Once the DNA construct has been created, amplified and purified out of the plasmid, it can be delivered into the eggs of the target organism.

Transgene Delivery

Several transgene delivery methods have been used but none give more than limited success. For this reason, special reporter genes have been used. These are simply genes that are very obvious if they have successfully been incorporated into the host genome—for instance, they code for proteins which cause the embryo to turn blue or fluoresce green. This makes it easier to compare the success of different delivery methods and develop the best technique for delivery of the true target gene.

Microinjection

Microinjection was first developed for use on mice and it is the most common transgene delivery method used for fish, though other methods seem to be more promising for molluscs. It simply involves using remote-control levers to operate a micropipette tipped with a very fine glass needle. Individual fertilised eggs are held steady by suction against a blunt ended tube and the micropipette needle is inserted into the egg.

A small volume of buffer (usually 1-2 nl) containing a high copy number (10^6-10^7 copies) of the transgene construct is then injected into the egg, the needle is removed and the egg incubated in the normal way. Some of the difficulties inherent in the microinjection of mouse eggs, such as working *in utero* with minute eggs, are removed when working with fish. Nevertheless, although fish eggs are abundant, large and easily accessible they do have characteristics that make microinjection difficult. Fish eggs are surrounded by a chorion that rapidly hardens after fertilisation, making it difficult to penetrate. Partly owing to the hardening of the chorion, it is also extremely difficult to identify the position of the chromosomes in fertilised eggs. Being able to locate the injected DNA as close as possible to the region where the chromosomes are located is an important criterion, because otherwise the integration of the novel DNA into the chromosomes of the GMO is far less likely. However, without special *in vivo* fluorescent staining and the use of a UV light source which enables precise localisation of the male and female pronuclei, injection of construct into the pronuclear region is bound to be rather hit and miss. Attempts have been made to remove the chorion from eggs before microinjection, but this turned out to be a slow and labour-intensive procedure which did not significantly improve success. In some species, the egg has a micropyle through which microinjection is feasible.

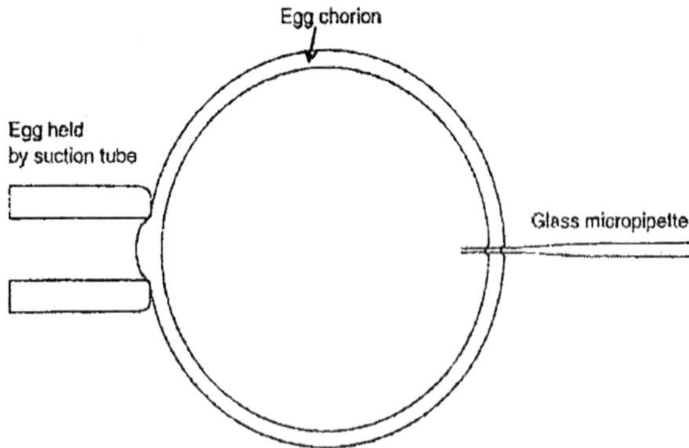

Fig. 1. Microinjection of DNA construct into a fish egg.

In contrast to mice, the time from spermatozoa activation to first cleavage in fish is quite short and early embryonic cell division is rapid, which means there is only a short window of opportunity for microinjection into the fertilised egg or early embryo. Even so, depending on the species of fish, an experienced operator with logistical support may be able to inject several hundred eggs before the cleavage divisions are too advanced. Ideally the transgene should be injected at the single cell stage to ensure integration in all the cells of the GMO. However, even when this is done, integration (if it happens at all) often only takes place after some cleavage divisions have occurred. This produces mosaic embryos where the transgene is present in some cells but not others. Mosaicism is a very common outcome of transgene delivery by the microinjection method.

In general, the injection method is rather inefficient. Injection can cause damage that affects embryonic survival and can result in quite high mortalities. Often fewer than 50% of eggs which have been microinjected express the transgenic DNA in early embryos and expression can decrease significantly after a few days. Usually fewer than 1% of microinjected salmonid eggs develop through to the adult stage with the transgene integrated into the genome, though better results can be obtained experimentally with smaller fish species such as zebrafish *(Danio rerio)* and medaka *(Oryzias latipes)*.

Electroporation

Electroporation is a method which was first developed for work with cells in tissue culture and it involves subjecting the cells to a short burst of electrical impulse. When cells are treated like this, their cell walls become temporarily much more porous and larger molecules than usual can pass through. Cells to be treated are suspended at high concentration in a solution of high copy number DNA construct and held in a 1-2 ml container with flat electrodes on each side. The transient current is passed and DNA construct molecules pass through the cell membrane. The electrical pulse can have a range of voltages and different rates of decay after the pulse and these are characteristics that are varied in order to produce optimum electroporation results. A major advantage of electroporation over microinjection is that there is no need to handle and manipulate eggs individually.

Electroporation has been tried on fish eggs, but the difficulty is that the eggs are quite large and have a chorion. Even after chorion removal, which is time consuming and therefore reduces the number of eggs that can be treated within the appropriate time window, the results have not encouraged further development. The stresses on embryos seem greater with this electrical method than the cell puncture used in the injection method, since embryo survival following electroporation is poor. While there are problems with fish eggs, the method has great potential for transgenesis in molluscs and successful experimental trials have been conducted with abalones and oysters. The eggs of most commercially important molluscan shellfish are rather small—less than 100 ljm in diameter—which makes it quite tricky to micromanipu-late individual eggs. On the other hand, there is no development of an impermeable layer following fertilisation and the egg cell wall remains unprotected by any kind of chorion. For these reasons electroporation has greater potential than microinjection for transgenesis in molluscan shellfish.

Sperm-mediated Transfer

When thought about, it's obvious. What is the one thing that is guaranteed to get into the egg during fertilisation and deliver a package of DNA directly to the female pronucleus? The spermatozoon, of course. It turns out that DNA binds readily to the outer coat of spermatozoa, so at first sight this looks like a perfect method of getting novel DNA into the egg. Encouraging though this prospect seems, it is evident that there are a number of pitfalls with this method. First, in most cases it is only a single spermatozoon that enters the egg and reaches the pronucleus. Therefore the number of copies of the DNA construct entering the egg or reaching the target is quite small when compared with the numbers of copies employed in electroporation or microinjection. Secondly, it is now clear that there are probably mechanisms in place in the host to prevent the ingress of foreign DNA by this process. Spermatozoa will often come into contact with extraneous DNA (from cellular debris or bacteria) before fertilisation so it is not surprising that mechanisms have evolved to prevent such DNA from integrating into the host genome. Attempts have been made to electroporate spermatozoa with DNA construct before fertilisation; although this approach has not been extensively explored, early work suggests that the difficulties of the methodology outweigh any benefits.

Biolistics

In plants, where there is a tough cell wall, experimenters have resorted to brute force to get foreign DNA into the cells. The method is called biological ballistics, or biolistics, and involves coating microscopic particles, usually of gold, with DNA construct and explosively firing these particles directly into the cell through the cell membrane. This method has been tried with fish, sea urchins and oysters and results suggest that viable transgenic embryos can be produced. However, there is currently no evidence that biolistics offers a more effective or more efficient method of producing transgenic fish than microinjection.

Viral Vectors

An alternative gene transfer method is to use as a vector a virus that has been rendered defective. The transgene is spliced into the defective virus that is nevertheless able to infect host cells and induce the replication of the transgene within those cells. However, there are obvious dangers involved in engineering any virus to be *too* good at getting into a wide range of species.

Lipofection

Synthetic lipid vesicles containing encapsulated DNA can be taken up directly by animal cells and this lipofection method has been tried with fish eggs. As with sperm-mediated transfer, initial expression of foreign DNA was followed by a rapid loss of expression suggesting that the novel DNA may have been destroyed by the host embryos.

All that these delivery methods do is to get the construct into the egg. It must then become incorporated into one or more of the chromosomes of the host. If it just remains floating around in the cytoplasm of cells, or as a fragment of DNA in the nucleus not attached to a chromosome, it will probably not be replicated and will be lost during cell division. Fortunately, a very small proportion of inserted construct becomes integrated spontaneously—which is what is currently relied upon in commercial transgenesis—but methods are being developed to increase transgene integration success.

Transgene Integration

A major difficulty with the integration of a transgene is targeting its position on a chromosome. Chromosomes are not homogeneous throughout their length but consist of heterochromatic and euchromatic regions, and genes in heterochromatic regions are seldom switched on. If the transgene gets incorporated in such regions then it is unlikely to be expressed due to this position effect. Furthermore, some research has shown that transgenes that had become successfully integrated and expressed in F0 and F1 fish were not expressed in later generations. Assuming the transgenes were still integrated, their expression must have been switched off in some way.

Three main experimental approaches are being used to try to get transgenes into the nucleus and to locate them more precisely on the chromosomes. First, many viruses have proteins containing special amino acid sequences that assist the viruses to get into the nucleus where they can replicate. These are called nuclear-localising sequences (NLS) and such sequences can be added to the medium containing the transgene construct to assist with entry into the nucleus.

Secondly, a pseudotyped retrovirus (murine leukemia retrovirus with one of its genes replaced by one from the vesicular stomatitis virus) that encodes a special inte-grase protein can be used. Once inside the cell the integrase protein is produced and this can assist with integration of the transgene DNA into the chromosome. However, there are significant risks attached to working with these viruses because they can infect any cell, including those of the human experimenter! Far better to try to isolate and utilise such integrase proteins directly and experiments to this end have been carried out.

The third approach involves harnessing the enzymes responsible for the insertion of the highly mobile DNA elements known as jumping genes or transposons. These transposases act in a fairly straightforward manner, cutting a DNA segment from one place in a chromosome and pasting it in at another on the same or a different chromosome. Natural transposases isolated from a number of organisms failed to work well, so a synthetic transposase has been manufactured and the result—the Sleeping Beauty (SB) transposon—effectively enhances integration (up to about 20-fold in zebrafish embryos).

None of these approaches is yet used commercially, but work in this area is increasing knowledge of the effects of positioning on gene function. Biologists still have very little information about what most of these genes actually do. However, using model species such as the zebrafish, transgenesis of experimentally-mutated genes provides the route by which more information about the functional genomics of fish will be elucidated.

Detecting Integration and Expression of the Transgene

The production of fish or shellfish GMOs by any method requires confirmation. To establish that the DNA construct is present in the GMO, that it has become integrated into the host genome and that the expected gene product is being expressed at the right level and in the right tissues for the particular commercial application for which the GMO was designed.

In order to detect the presence of the transgene in an early embryo, a simple method is to carry out dot blots. DNA is extracted from putative transgenic embryos or larvae and the extracts are spotted onto a nylon membrane in a simple grid pattern. The DNA is heat-treated to denature it and cause it to adhere to the membrane, and the membrane is then hybridised with a radiolabelled probe of a DNA sequence complementary to some part of the transgene.

Dot blots and PCR help to identify those individuals in which the transgene is present, but they do not demonstrate that the DNA has actually been incorporated into the chromosomes. In fact, because only a tiny proportion of the millions of transgene copies used per egg are ever integrated into the genome, there is bound to be extensive non-incorporated transgene. In order to provide evidence of transgene incorporation, a method involving restriction enzyme incubation and Southern blotting has been used. The extracted DNA from the putative transgenic organism is incubated with a selected suite of restriction endonucleases that cut the DNA into fragments of different lengths. These fragments are then size-sorted on an electrophoretic gel and Southern blotted, with the resulting membrane being probed and visualised in the usual way. If the transgene has not become incorporated into the chromosomes then its length (and therefore position on the gel) should be the same as before it was introduced into the organism. However, where the transgene has become incorporated into the genome, it will occur in fragments of greater length than the original transgene. Not only can those individuals which have the transgene be identified using this method, but also integration of the transgene into the chromosomal material can be indicated. Unfortunately, there remains room for error here because of the vast numbers of unintegrated transgenes, many of which can recombine in unknown ways. Recombined transgenes will be longer than single transgenes and could lead to individuals, in which the transgene has not become incorporated, being wrongly identified as GMOs following blotting.

Ultimately the surest method of confirming the integration of a transgene into an individual is to demonstrate that it has passed the transgene on to its offspring in a true Mendelian fashion. Using the simplest assumption of the transgene being integrated into a single chromosome, then only half of all gametes produced by the transgenic individual (the F0 generation) will contain the transgene. Therefore the expectation is that only at most 50% of the F1 generation will contain a copy of the transgene.

How can discover if the transgene is being expressed? One method of identifying the protein is to use an immunological approach where the presence of a particular protein can be identified by raising antibodies against it. Alternatively, of course, can simply measure the phe-notypic consequences of the transgene in the GMO. For example, with the use of an antifreeze protein, GMOs can be tested for cold tolerance. Similarly, if a growth hormone gene has been used, can assess whether the transgenic group grow significantly faster than non-GMO contro (Table 2).

Table 2. Increase in growth rates in transgenic fish relative to controls over a single generation when using growth hormone genes from different sources

Species	*Gene*	*% growth increase*
Carp	Trout GH	50-90%
Tilapia	Tilapia GH	50-90%
	Salmon GH + ocean pout anti-freeze promoter	200-400%
Salmon	Chinook salmon GH + ocean pout anti-freeze promoter	>1000%

GH = growth hormone.

Here will consider the results of an experiment by R.H. Devlin and colleagues on genetically modified coho salmon *(Oncorhynchus kisutch)*. First, coho salmon eggs were microinjected with an 'all fish' gene construct consisting of the chinook salmon growth hormone gene and the ocean pout antifreeze protein promoter. Some of the transgenic salmon produced were male and matured at 2 years old. Five of these fish were mated to wild-type coho salmon with the expectation that about 50% of the offspring should contain the transgene. Identification was based on the PCR amplification of a specific segment of the ocean pout promoter sequence which produced a fragment only when the DNA analysed came from a transgenic fish. It was noted that the post-larval fish—alevins—were of two different colours. The majority had the normal brown coloration but some were much paler with a distinct greenish hue. PCR tests showed that the green alevins were transgenic, while the brown ones were not. This unexpected phenotypic marker of transgenic coho salmon alevins provided a useful identifier without having to carry out PCR analysis.

The greenish coloration was not a permanent marker because after first feeding, the colour of transgenics, although paler, gradually became much less distinguishable from the colour of non-transgenic fish. Analysis of the offspring from the five crosses showed that the percentages of transgenic offspring were always much less than 50%, ranging from zero to 19%. This is a common finding in F1 transgenics and probably indicates that the F0 generation were mosaics with the mosaicism extending into the germ cells. Either that, or the transgene may have been subject to loss during meiosis because of being in the haploid rather than the diploid state.

At the pre-feeding alevin stage the transgenic fish were significantly heavier and longer than the non-transgenics, providing evidence for the enhanced activity of growth hormone in the fish. However, associated with the enhanced size was evidence of morphological abnormality, even at this early stage.

Transgenic fish had larger heads than non-transgenics and this abnormality became more pronounced as the F1 fish got older. Some severely abnormal fish had difficulty feeding due to excessive jaw growth and had problems ventilating their gills because of the extent of operculum overgrowth. Many of these F1 fish died. Similar abnormalities had been apparent in the F0 generation of transgenics.

The results of this early experiment illustrate some of the potential problems arising from transgenesis, though since these trials there have been considerable improvements that reduce this type of growth abnormality in transgenic fish. Nevertheless, much more research is clearly required before transgenic aquacultured organisms can be expected to receive consumer acceptance, and thus commercial success. Unfortunately, the potential commercial value of genetic engineering has meant that most of the near-market research has been, and will continue to be, conducted by private companies that employ their own researchers and have no need to publish their findings in peer-reviewed journals. Concern and suspicion about the integrity of such research activity will inevitably arise because it is paid for by the companies themselves.

Performance of Transgenic Fish

Growth

Positive biological effects have been obtained by transferring transgenes to fish in some, but not all, cases. Initially, this research focused on the transfer of foreign GH gene constructs into fish. Due to the lack of available piscine gene sequences, transgenic fish research in the mid-1980s employed existing mammalian GH gene constructs, and growth enhancement was reported for some fish species examined. Mammalian gene constructs mouse metal-lothionein/rat growth hormone (mMT/rGH) failed to affect the growth of salmonids, despite the fact that salmonids are very responsive to growth stimulation by exogenously administered mammalian GH protein. Gene constructs containing fish GH sequences driven by non-piscine promoters elicited growth enhancement in transgenic carp, catfish, zebra fish and tilapia. Growth stimulatory effects observed with the above constructs have ranged from no effect to twofold increases in weight relative to controls, and provided the first convincing data demonstrating that growth enhancement in fish can be achieved by transgenesis.

However, subsequent experiments demonstrated that growth can be enhanced through transgenesis from 10% up to an incredible 30-fold in some conditions. Like other breeding programmes, sometimes no enhancement is obtained. The results are basically consistent. Several species, including loach, common carp, crucian carp, Atlantic salmon, channel catfish, tilapia, medaka and northern pike, containing either human, bovine or salmonid GH genes are reported to grow 10-80% faster than non-transgenic fish in aquacul-ture conditions, if the proper promoters are utilized. Chen *et al.* have shown integration of hGHg in loach and significant increases in the length and weight of these fish.Du *et al.* used an all-fish GH gene construct to make transgenic Atlantic salmon.

Similar results have been obtained for transgenic *Oreochromis niloticus* possessing one copy of an eel (ocean pout) promoter-chinook salmon GH fusion: they grew 2.5- to fourfold faster and converted feed 20% better than their non-transgenic siblings. At 7 months, the mean body weight of transgenic tilapia was 653 g compared with 260 g for non-transgenic siblings. These were heterozygous lines of GH transgenic Nile tilapia, and the accelerated growth was obtained in the F1 and F_2 generations. However, F_1 fish transgenic for a construct consisting of a sockeye salmon MT promoter spliced to a sockeye salmon GH gene exhibited no growth enhancement, although salmon transgenic for this construct show greatly enhanced growth. The growth-enhanced transgenic lines of Nile tilapia were strongly positive for the salmon GH in their serum, whereas the non-growth-enhanced lines were negative. Attempts to induce expression from the MT promoter by exposing fish to increased levels of zinc failed. Preliminary results indicated that homozygous transgenic Nile tilapia produced from the ocean-pout antifreeze-chinook salmon GH construct have growth similar to that of the hemizygous transgenics.

Insertion of other GH constructs into tilapia have also yielded positive results, but not as dramatic as those with the salmon GH constructs. Two possible explanations for the difference in results are that the type of construct and the type of tilapia studied were different. Introduction of a CMV-tiGH construct into a hybrid *O. hornorum* resulted in a 60-80% growth acceleration depending on the culture conditions. Different patterns and levels of ectopic expression of tiGH and tilapia insulin-like growth factor (IGF) were detected in organs of four lines of transgenic tilapia by RNA or protein analysis. The two lines with lower ectopic tiGH mRNA levels were the only ones that exhibited growth acceleration, suggesting that the expression of ectopic tiGH promoted growth only at low expression levels. Higher ectopic tiGH levels resulted in a low condition factor. Overexpression of tiGH had no positive or negative effects, similar to the result observed in transgenic GH pigs but opposite to what was observed in transgenic GH salmon exhibiting hyperlevels of GH expression.

One of the most thorough studies of GH gene transfer is that of the transfer of the rtGH cDNA driven by the RSV-LTR promoter into channel catfish and common carp. Transgenic individuals of some families of carp and catfish grow 20-60%

faster than their non-transgenic full siblings, but in some families no differences exist. Differences in genetic background, epistasis, copy number of the foreign gene, insertion site and level of expression are logical explanations for these results, which also illustrate the fact that a combination of traditional breeding programmes, such as selection, along with gene transfer will probably be necessary to develop the best genotypes for aquaculture.

Zhang *et al.* have shown the gene transfer, expression and inheritance of rtGH cDNA in the common carp. Progeny that inherited the transgene from transgenic common carp possessing the RSV-rtGH cDNA grew 20-40% faster than full siblings that did not inherit the gene. Thirty to fifty per cent of the transgenic progeny were larger than the largest non-transgenic sibling. The coefficient of variation for body weight was similar for transgenic and non-transgenic siblings, indicating that the population distribution was the same for body weight. The percentage of individuals that were deformed was not different between transgenic and non-transgenic progeny.

Transgenic common carp and silver crucian carp possessing MThGHg grew 11 and 78% faster than non-transgenic controls, respectively. The different response of the two species can be explained by the variable expression of individuals, differing insertion sites and therefore differing regulation and expression, or small sample sizes.

The transgenic common and silver crucian carp had a coefficient of variation twice that of non-transgenic controls. This is in contrast to the results of Zhang *et al.*, and may be due to the less consistent expression of the transgenic carp in the experiments of Chen *et al.* All transgenic carp evaluated that possessed RSV-rainbow trout cDNA expressed recombinant GH, while only 50% of the transgenic carp that possessed MThGHg expressed recombinant growth hormone. Variable expression should lead to variable growth. Loach containing MThGHg and northern pike containing RSV-bovine GH gene also exhibited more variable growth than controls. Combining selection with genetic engineering procedures might reduce this variation.

Transgenic salmon illustrate the most dramatic results obtained in fish genetic engineering. The GH gene constructs utilized were comprised entirely of piscine gene sequences using either an ocean-pout antifreeze promoter (opAFP) driving a chinook salmon GH cDNA, or a sock-eye salmon MT promoter driving the full-length sockeye GH1 gene. When introduced into salmonids, these gene constructs elevated circulating GH levels by as much as 40-fold, resulted in up to five- to 30-fold increase in weight after 1 year of growth, and allowed precocious development of physiological capabilities necessary for marine survival (smoltification). The largest of these P1 transgenics were mated and produced offspring with extraordinary growth.

The extraordinary accelerated growth was obtained in a number of salmonid species. The opAFP-chinook salmon GH gene construct accelerated growth in coho salmon by ten to 30 times. Parr-smolt transformation occurred 6 months early in the transgenic fish compared to the control fish. Results varied among species and families and might be related to different gene constructs, coding regions, chromosome positions and copy numbers. Insertion of opAFP-chinook salmon GH1 cDNA increased growth 3.2-fold in rainbow trout, whereas the sockeye MTB-sockeye GH1 accelerated growth tenfold. The opAFP-chinook salmon GH1 cDNA construct improved growth tenfold in cutthroat trout, *Oncorhynchus clarki,* and 6.2-fold in chinook salmon. Growth at 5 months was better than growth at 15 months, again illustrating that, on a relative basis, genetic improvement for growth is usually more impressive at younger ages than at older ages. Condition factor, K, was lower for transgenic fish because length changed more rapidly than weight. Some families that had 30-fold increased growth exhibited acromegaly in the jaw, skull and opercular area and, by 15 months, the growth of these fish slowed and they died. As was seen with transgenic common carp and channel catfish, the effect of GH gene insertion was variable among families, and multiple insertion sites and multiple copies of the gene were observed. Transgenic rainbow trout experienced early maturation at 2 years of age, but in the same season as the controls.

Sockeye MTB-sockeye GH cDNA1 introduced into coho salmon increased growth 11- to 37-fold and increased GH expression by 40-fold in cold temperatures, when GH expression is normally low. Results with Atlantic salmon are not quite as impressive as with coho salmon. Transgenic Atlantic salmon containing the opAFP-chinook salmon GH cDNA1 gene construct had a three- to sixfold accelerated growth rate compared to non-transgenic salmon. Insertion of sockeye MTB-sockeye GH cDNA1 produced a similar result, fivefold growth enhancement.

Prior to first feeding, the transgenic progeny were found to be 21.2% heavier and 11.9% longer than their non-transgenic siblings, suggesting that the expression of GH in early development can influence the rate or efficiency of conversion of yolk energy reserves. Magnification effects can explain some of the growth differences between transgenic and control salmon; however, specific growth rates of the transgenic coho salmon were approximately 2.7-fold higher than those of older non-transgenic animals of similar size and 1.7-fold higher than those of their non-transgenic siblings, indicating that the transgenic salmon are growing at a faster rate at numerous sizes and life stages. GH levels were increased dramatically (19.3- to 32.1-fold) relative to size-control salmon, but IGF-I levels were only modestly affected, being slightly enhanced in one experiment and slightly reduced in another. Insulin levels in transgenic animals did not differ from same-size controls, but were higher than in non-transgenic siblings, and thyroxine levels in transgenic salmon were intermediate between levels found in size and age controls.

Growth enhancement varies greatly among different transgenic fish systems, and salmonids in particular have shown the greatest response to stimulation to date. Several potential explanations exist that indicate that it may be difficult to duplicate these results in other fish species.

It is possible that completely homologous gene constructs, derived only from the same species or from piscine sources, such as those used successfully in salmonids, are expressed in fish more efficiently than are gene constructs derived from other vertebrates. While this probably plays a role in efficient expression, it is also very likely that the differences in growth response observed in different transgenic systems are due to the vastly different physiologies and life-history characteristics that exist among the fish species examined.

The biology of salmon and their unique physiological adaptations probably play an important role in the dramatic growth enhancement observed in transgenic individuals. Growth in salmonids is normally relatively slow throughout the year, and is extremely low when water temperatures are low and food resources in nature are scarce. This low growth rate appears to be controlled at least in part by the level of circulating GH, and can be dramatically stimulated with exogenous GH protein and sufficient food. The dramatic growth stimulation observed in transgenic salmonids may arise partially from the seasonal deregulation of GH expression to allow high growth rates during winter months when control salmon have very slow growth rates. This accelerated winter growth may also give the transgenic fish a large advantage that can later be magnified. Additionally, salmonids are anadromous, and accelerated growth in transgenics allows them to reach a size at which they smolt earlier than controls. Growth in the smolt stage is naturally increased, providing transgenic individuals with a further advantage to increase the relative growth difference from controls.

Non-salmonid fish species often display more rapid growth, and consequently may be much more difficult to stimulate by expression of GH in transgenic organisms. These high growth rates occur naturally in some species such as tilapia, whereas others have been enhanced through genetic selection and through many years of domestication. Strains or species that have been selected to near maximal growth rates presumably have had many of their metabolic and physiological processes optimized, and might be expected to be more difficult to stimulate by a single factor such as GH.

Domestication is also important in transgenic growth responses, and Devlin *et al.* observed that salmonid GH gene constructs that have a dramatic effect on growth in wild rainbow trout strains (with naturally low growth rates) have little or no effect in strains where growth rate has been enhanced by selection over many years. P1 and F1 rainbow trout derived from a slow-growing wild strain and containing salmon metallothionein growth hormone (OnMTGH1) grew 17-fold faster than controls. Transgenic males and females eventually reached 8.2 and 14.2 kg, compared with 220 and 171g for non-transgenic males and females, respectively. These wild transgenic rainbow trout grew no faster than a fast-growing non-transgenic, domestic rainbow trout. This is indicative of the tremendous progress that domestication and selection have had on some aquaculture species and implies a possible ten- to 20-fold improvement of growth through selection.

Introduction of the transgene into the domesticated P1 only increased growth by 4.4%; however, replication was extremely low, and fast-growing families could have been missed. These data indicate that the effects of selection and transgenesis are similar, but not additive in any way. Supportive of the conclusions reached by Devlin *et al.*, rainbow trout treated with exogenous GH protein exhibited similar effects to those in the wild and domestic transgenic rainbow trout. Body weight of untreated wild-strain rainbow trout increased at a modest specific growth rate of 0.68%/day and untreated domestic rainbow trout grew at 2.18%/day; however, specific growth rates of hormone-treated wild-strain trout were enhanced 2.7-fold, similar to the untreated domestic fish, and domestic rainbow trout had an increase of only 9%. Consistent with these observations, the dramatically growth-responsive salmon previously observed were also derived from wild strains. Apparently, slow-growing wild strains can benefit much more from GH insertion than fish that already have growth enhancement from selective breeding.

Similarly, dramatic growth stimulation in the mammalian system using GH transgenes has been observed in mice, but not in selected mice and domestic livestock that have had many centuries of genetic selection. In these domesticated and selected lines, the capacity for further growth improvement by GH may now be restricted by limitations in other physiological pathways, and other methods, including traditional selective breeding methods, may yield the greatest gains. For aquacultural species, which have a much shorter history of domestication and selection, future genetic improvement will probably be accomplished by utilizing a combination of both approaches simultaneously.

In contrast and in comparison, GH transgenic catfish derived from domesticated and selectively bred strains exhibit only a moderate growth enhancement (41%). Thus it appears that growth of wild fish can apparently be improved in one or two generations with the insertion of GH genes to the extent that would take many generations of selective breeding to achieve.

However, additional data on transgenic rainbow trout refute this hypothesis of the effect of wild and domestic genetic backgrounds on response to GH transgene insertion. When OnMTGH1 was transferred to another wild rainbow trout strain, F77, growth was enhanced sevenfold, which was almost a fourfold greater growth than that observed in a non-transgenic domestic rainbow trout. In this case, the wild transgenic rainbow trout is actually superior to the domestic selected strain, indicating that genetic engineering can have a greater effect than, rather than an equivalent effect to, domestication and selection. Perhaps strain effects in general, epistasis and genetic background are more significant in regard to affecting transgene response, rather than the domestic or wild nature of the fish. When F77 was crossbred with the domestic strain, growth of the crossbreed was intermediate to the parent strains, a typical result. However, the transgenic wild **X** domestic crossbreed was by far the largest genotype, 18 times larger than the non-transgenic wild parent, 13 times larger than the non-transgenic wild **X** domestic crossbreed, nine times larger than the non-transgenic domestic parent and more than 2.5 times larger than the wild *F77* transgenic parent.

The combined effects of transgenesis and crossbreeding had a much greater growth enhancement than crossbreeding or transgenesis alone. Additionally, a transgenic with 50% of its heritage from domestic sources was much larger than a wild transgenic, so great response from some domestic genotypes is possible.

Cold Tolerance

Most efforts in transgenic fish have been devoted to growth enhancement, although there are reports of improvement in cold and disease resistance. Early research also involved the transfer of the antifreeze protein gene of the winter flounder. The primary purpose of this research was to produce salmon that could be farmed under Arctic conditions, but expression levels obtained have been inadequate for increasing the cold tolerance of salmon. However, preliminary results with goldfish show some promise for increasing survival within the normal cold temperature range. Initial thrusts at improving cold tolerance via gene transfer have been minimally successful.

Disease Resistance

Momentum is being gained in transgenic enhancement of disease resistance. Expression of viral coat protein genes or antisense expression of viral early genes may improve virus resistance, although bacterial diseases are often greater threats to the major aquaculture species, and bacterial disease resistance may be easier to genetically engineer than resistance for diseases caused by other classes of pathogens.

One potential mechanism for improving disease resistance is the production of transgenic aquatic organisms containing lytic peptide genes. A great deal of information is available concerning antibacterial peptides; the number of structural families into which they fall is very large, it seems likely that they occur ubiquitously and organisms containing these genes should exhibit enhanced disease resistance. The earliest discovered and most thoroughly studied antibacterial peptides are the cecropins, small cationic peptides found originally in the moth *Hyalophora cecropia*. Antimicrobial peptides other than cecropin have been identified in many invertebrates and vertebrates.

Cecropins are initially translated into precursors of 62-64 amino acid residues, and are then processed intracellularly into mature peptides of 35-37 amino acid residues. The unique structural features of cecropins and other antimicrobial peptides allows them to readily incorporate into cellular membranes of bacteria, fungi and parasites, resulting in the formation of pores on the membrane, leading to the inevitable death of pro- and eukaryotic pathogens. Cecropin analogues can be designed and synthesized that are as effective as, or even more potent than, the native compounds against animal and plant bacterial pathogens and protozoa. Cecropin genes and their analogues have been used to produce transgenic plants, such as potato and tobacco, with increased resistance to infection by bacterial or fungal pathogens.

Several studies have demonstrated the *in vitro* effectiveness of cecropins. Cecropins are of potential benefit in aquaculture because they show a broad spectrum of activity against Gram-negative bacteria, which include most of the major bacterial pathogens of ictalurid catfish, they are non-toxic to eukaryotic cells, they are found in mammals, as described for the pig, as well as insects, and there is an extensive literature on the physicochemical properties and mode of action of cecropins, which will facilitate the redesign of cecropin derivatives, if necessary. Passively administered cecropin derivatives can confer some protection against infection with *Edwardsiella ictaluri,* a major pathogen of cultured catfish. Chiou *et al.* examined the *in vitro* effectiveness of native cecropin B and a synthetic analogue, CF17, for killing several fish viral pathogens, infectious haematopoietic necrosis virus (IHNV), viral haemorrhagic septicaemia virus (VHSV), snakehead rhabdovirus (SHRV) and infectious pancreatic necrosis virus (IPNV). When these peptides and viruses were co-incubated, the viral titres yielded in fish cells were reduced from several- to 104-fold. Direct disruption of the viral envelope and disintegration of the viral capsids may be the explanation for the inhibition of viral replication by the peptides. These antimicrobial peptides were more effective on enveloped viruses than on non-enveloped viruses, and may act at different stages during viral infection to inhibit viral replication. Three mechanisms—direct inactivation of viral particles by perturbing the lipid bilayers of the viral envelopes, prevention of viral penetration into the host cell by inhibiting viral-cellular membrane fusion and inhibition of viral replication in infected cells by suppressing viral gene expression—have been proposed to explain the antiviral action. The same antimicrobial peptides act on different viruses through different mechanisms.

The *in vitro* studies indicate that transgen-esis utilizing cecropin constructs should improve disease resistance in aquatic organisms. Bacterial disease resistance may be improved up to three- to fourfold through gene transfer. Insertion of the lytic peptide cecropin-B construct enhanced resistance to bacterial diseases two- to fourfold in channel catfish. There was no pleiotropic effect on growth. P_1 transgenic catfish containing the cecropin-B construct were spawned and the transgene was transmitted to the F1 generation. Transgenic and non-transgenic full siblings containing the cecropin-B construct were challenged in tanks with *E. ictaluri.* Both genotypes experienced mortality, but the survival of the transgenic individuals was twice that of the controls. Transgenic channel catfish containing the preprocecropin-B construct and their full-sibling controls experienced a natural epizootic of columnaris, *Flavobacterium columnare.* No cecropin-transgenic fish were among the mortalities, and only control fish died. Both transgenic and control individuals were among the survivors. In this case, the transgene appears to have imparted complete resistance. Catfish transgenic for cecropin constructs show enhanced resistance to both deliberate and natural challenge with pathogenic bacteria.

Similar results were obtained for cecropin-transgenic medaka. F_2 transgenic medaka from different families and controls were challenged with *Pseudomonas fluorescens* and *Vibrio anguillarum,* killing about 40% of the control fish by both pathogens, but only 0-10% of the F_2 transgenic fish were killed by *P. fluorescens* and about 10-30% by *V. anguillarum.* When challenged with *P. fluorescens,* zero mortality was found in one transgenic fish family carrying preprocecropin B and two families with porcine cecropin P1 and 0-10% cumulative mortality for five transgenic families with procecropin B and two families with cecropin B. When challenged with *V. anguillarum,* the cumulative mortality was 40% for non-transgenic control medaka, 20% in one transgenic family carrying preprocecropin B, between 20 and 30% in three transgenic families with procecropin B and 10% in one family with porcine cecropin P1. RT-PCR analysis confirmed that transgenic fish from most of the F_2 families expressed cecropin transgenes except those in three F_2 families.

Other antimicrobial peptides have shown promise for the protection of fish against pathogens and might be utilized with transgenic approaches. A cecropin-melittin hybrid peptide (CEME) or pleurocidin amide, a C-terminally amidated form of the natural flounder peptide, was delivered continuously using miniosmotic pumps placed in the peritoneal cavity, followed 12 days after pump implantation with intraperitoneal injections, into juvenile coho salmon infected with *V. anguillarum*, the causative agent of vibriosis (Jia *et al.*, 2000). Juvenile coho salmon that received 200 lig of CEME/day had lower mortality (13%) than the control groups (50-58%), as well as fish that received pleurocidin amide at 250 lag/day which had lower mortality (5%) than control groups (67-75%).

Transgenic Production of Pharmaceuticals

Another use of transgenic organisms is as biological factories to produce highly valuable pharmaceutical proteins. With the advent of diseases such as acquired immune deficiency syndrome (AIDS) and hepatitis, it became even more important to develop alternatives for extracting compounds such as blood-clotting factors from human blood. Transgenic goats, cows and other livestock have been developed that produce valuable biomedical products. Usually, the transgenically produced protein is extracted from the milk. The potential exists to apply this same concept in aquatic organisms, and recently transgenic tilapia have been developed that express human factor VII (N. Maclean, personal communication) and transgenic rainbow trout developed expressing human proteins.

Gene Knockout Technology

Theoretically inactivation of a gene can be accomplished by knockout of the gene by replacing the original gene with a mutated copy of the gene or by disruption of gene expression using the antisense approach or ribozyme technology. Although the latter two approaches are currently feasible, the knockout approach is the ultimate method for gene inactivation because it will eliminate the gene products completely. Disadvantages of the antisense and ribozyme approaches are potential problems with position effect after integration and transgene inactivation in F_1 or later generations due to methylation of the transgenes.

Gene knockout has not yet been accomplished with aquatic organisms. Gene knockout efficiency would probably be low since the selection methodology has not been developed to screen for the proper genotypes. A second approach using pluripotent embryonic stem (ES) cells or primordial germ cells has been employed for the production of knockout transgenics in mice. Because ES cells and primordial germ cells are pluripotent and contribute to the germ-cell lineage when transplanted into host embryos, these cells, if transformed, can serve as a vector for the introduction of foreign DNA into the germ line of the organism. *In vivo* pluripotency has been demonstrated by transmission of ES-cell genotypes to chimeric offspring in rabbit and pig and recently in fish. Additionally, pluripotency of embryo-derived cells has been demonstrated by bovine conceptus development and the birth of live lambs after nuclear transfer.

The use of cultured cells as a vector for the production of transgenic aquatic organisms is advantageous because it eliminates the problem of mosaicism, since *in vitro* selection follows transfer of foreign DNA into the cultured cells to isolate cell clones that have stably integrated and properly expressed the transgene. Additionally, homologous recombination can be used to inactivate or replace endogenous genes by targeted insertion, which has resulted in the production of knockout mice for the examination of specific gene function without changing any other genomic or physiological condition.

Homologous recombination activity has been observed in zebra-fish embryos, indicating the potential for knockout and knock-in technology in fish. Chen *et al.* then developed homologous recombination vectors and positive-negative selection procedures for fish cells. The positive-negative selection procedures were functional, but the gene targeting was not achieved.

Despite the advantages of using primordial germ cells and ES cells for transgenic research, this technology has not been fully developed for use with aquatic organisms. The primary disadvantage of this technology is that the founder knockout stock would be derived from an extremely small genetic base. This could be corrected by crossbreeding, but it would take two generations to develop an outbred population that was homozygous for the knockout construct.

The first step for developing this technology is the isolation of ES cells form aquatic organisms, and some initial work has been conducted with zebra fish. Pluripotent embryonic cells exhibiting *in vitro* characteristics of ES cells have been cultured and they contributed to tissues derived from all three germ layers when introduced into developing host embryos. These cultures were obtained from blastulae. These zebra-fish embryonic cells exhibited a morphology similar to that of ES cells, and were induced to differentiate into multiple cell types in culture, including melanocytes and muscle and nerve cells. Since this earlier work, ES-like cell lines have been developed from medaka and sea bream. The zebra-fish cells stained positive for alkaline phosphatase activity, an indicator of pluripotency

The medaka cell line, MES1, has shown pleuropotency retains the aneudiploid genotype and forms viable chimeras when injected into blastulae that later contribute to all three germ layers. Genetic background has a key role in establishing these ES cell lines.

However, the ES-like cells have yet to contribute to the germ-line cells of a host embryo. One explanation for the absence of germ-line transmission is that fish cells are committed to the germ line very early during embryogenesis, prior to the blastula stage from which the ES cells were derived and injected. Another option would be to inject cells before commitment to the germ line has occurred, but injection of cells prior to blastulation is technically difficult. A more logical option may be to utilize primordial germ cells, the embryonic precursors of germ cells.

In vitro selection can also be conducted on primordial germ cells to establish cell lines containing genomes with specific genes knocked out. The strategy of knockout is to produce transgenic fish that are homozygous for null genes, which have negative effects on commercially important traits. One difficulty is identifying and marking primordial germ cells. Alkaline phosphatase activity can be used as markers for germ cells, and vasa, a DEAE box protein family and a homologue of *Drosophila* vasa, has been cloned from mouse and *Xenopus* and has been used for marking germ cells. Vasa mRNA is visible in embryos at the four-cell stage and, by mid-blastula, four distinct small groups of primordial germ cells can be visualized by *in situ* hybridization using a vasa probe. At the early somite stage the primordial germ cells should aggregate on the ventral side of the embryo near the interface with the developing yolk-sac.

An alternative method for identifying primordial germ cells is to examine the cells for their ability to produce well-differentiated embryoid bodies similar to pluripotent ES cells. Primordial germ cells are homologously transfected with the gene-targeting vectors and selected using drugs such as G418 (neomycin is used in the vector). Prior to reintroduction into host embryos by microinjection, transformed cells need to be karyotyped to ensure that they still contain the full chromosome complement.

Theoretically, the transformation of the embryos can be accomplished by combining androgenesis and microinjection or by traditional nuclear transfer. In the case of the androgenetic approach, eggs are irradiated with gamma rays or UV rays to enucleate or destroy the maternal DNA. The irradiated eggs are activated with UV-irradiated sperm so that there is no paternal contribution. Transformed primordial germ cells are microinjected into the enucleated eggs for the production of transgenic fish.

Knockout technology has been utilized to develop transgenic mice with greatly altered growth and body composition. Disruption of the myostatin gene, GDF-8, resulted in GDF-8-null mice with a 262% increase in muscle, a 25% increase in growth rate and a 27% decrease in fat percentage compared with controls. Homozygous mutants of the knockout mice were viable and fertile.

Additional techniques for gene silencing or knockout exist. Not all of them can be used for transgenesis. Some of the techniques do not have an effect on the germ line, but can be used to study gene expression, perhaps leading to future gene manipulations.

The insertion of a transgene can sometimes have a paradoxical effect and actually silence rather than enhance gene expression, apparently due to the effects of the mRNA that is generated. Another technique for post-transcriptional gene silencing is the utilization of single-stranded RNA antisense constructs; however, Fire *et al.* found that double-stranded RNA was more effective at gene silencing—RNA interference—than single-stranded RNA, and some earlier success of single-stranded RNA was actually due to contamination with double-stranded RNA. For RNA interference to work, an organism apparently has to have specific genes present. RNA-dependent RNA polymerases may need to be present, and this phenomenon may be associated with the evolution of defence mechanisms against RNA viruses. RNA interference can be accomplished in vertebrates. Both double-stranded RNA and antisense RNA were effective in disrupting the expression of GFP in transgenic zebra fish.

Utilization of antisense oligos is another alternative for gene silencing. Antisense oligos need to possess the following traits: they must achieve efficacy in the cell at reasonable concentrations, should inhibit the target sequences without attacking other sequences, need to be stable extracellularly and within the cell, and must be deliverable to the cellular compartments containing the target RNA. Antisense oligos need to be as structurally similar to DNA as possible without coming under attack by nucleases.

The original antisense oligos were made from natural genetic material, with artificial crosslinking moieties added to irreversibly bind the antisense oligos to their target RNA, but these first-generation oligos were usually degraded naturally. Methylphosphonate-linked DNA oligos were then designed, which resisted enzymatic degradation, but they had poor efficacy and poor aqueous solubility. Phosphorothionate-linked DNAs (S-DNAs) were next developed, resulting in greater efficacy and solubility; however, they had the disadvantages of a narrow range of efficacious concentrations and various interactions with non-target proteins. Morpholines have overcome most of these problems. They contain 6-membered morpholine backbone moieties, joined by non-ionic phosphorodiamidate intersubunit linkages. Morpholines are designed for their excellent RNA-binding ability and their specificity. The design and construction of morpholines are quite complex. This is a useful tool for studying gene expression, but production of a morpholine transgenic fish is probably extremely difficult to accomplish.

Targeted gene knock-down or knockout has been accomplished in zebra fish. Antisense, morpholine-modified oligonucleotides (mor-pholinos) blocked the ubiquitous expression of the GFP transgene in transgenic zebra fish. This approach also generated phenocopies of the mutations of the genes no tail, chordin, one-eyed-pinhead, nacre and sparse and knocked out or reduced expression of the uroporphyrinogen decarboxylase, sonic hedgehog and tiggy-winkle hedgehog.

Casanova mutant zebra-fish embryos lack endoderm and develop cardia bifidia, and Dickmeis *et al.* were able to duplicate this condition by knocking out the HMG box-containing gene, 10J3, with morpholine oligonucleotides.

Manipulating and transplanting nuclei to accomplish gene knockout is difficult and has yet to produce a transgenic aquatic organism. Manipulating blastomeres to either add or knock out genes may be a future possibility. Takeuchi *et al.* were able to produce germ-line chimeras of rainbow trout. Orange mutant embryos were used for donor cells to wild-type embryo donors, and cells were transplanted utilizing microinjection. Mid-blastula and early blastula were optimum developmental times for donor and host embryos. Survival of manipulated embryos was 12%, and 50% of these fish exhibited mutant coloration. When mated with wild-type individuals, 32% of the chimeras transmitted the orange colour to progeny at a rate of 0.4—14%; therefore the orange mutant must be a dominant mutant.

Genethics

Genetic engineering and cloning have opened a can of ethical worms that have wriggled into all aspects of modern society. Many of the questions raised by these technologies are of a philosophical or ethical nature. These questions are quite properly the remit of all in society—individual scientists are as entitled to their own views on these questions as is any other member of the public. The fact that they are scientists does not give them any advantage on such philosophical matters or ethical questions.

On the other hand, there are some important questions that scientists must ask about the research they are doing in the field of genetic engineering. First must consider what is being classed as a GMO. Is a tetraploid oyster a GMO? Such an organism is artificially engineered (indeed, even patented) and not present in the wild and in that sense is unnatural. However, there is no genetic material from another species present in its genome, only two extra sets of its own chromosomes. In the view of most geneticists, tetraploids, and other products of ploidy manipulation such as triploids, androgens and gynogens, should not therefore be classed as GMOs.

How about the artificial hybridisation of two species of closely related fish that nevertheless are geographically isolated from one another? In the normal course of events (on the human time-scale) these two fish species would never meet. Here, in such hybrids, do have genetic material from two different species in the same organism: surely these are genetically modified organisms? Well, again, most geneticists would argue that this kind of event has been happening over evolutionary time in a natural way and that may just be speeding-up a natural process as indeed is being done in ordinary selective breeding. So such hybrids are not GMOs.

Finally, if consider a fish that carries in its genome an extra growth gene from another vertebrate species together with a promoter from a third species, then clearly this is what mean by a genetically modified organism. Although such a transfer of genetic material from one species to a totally unrelated one would seem at first sight to be unnatural, it is nevertheless known that in the wild, pieces of genetic material have been, and are still being, transferred occasionally between phylogenet-ically distinct organisms. Surprisingly, this is a natural process.

The most important practical questions to do with genetic engineering are questions about risk. What is the risk of the escape of genetically modified organisms into the wild? If, in spite of all efforts to reduce the chance of escape, an accident occurs and some GMOs—or their gametes—escape, what risk does that pose to (a) humans, (b) non-GMOs of the same species and (c) the environment? The balance of environmental risk against the actual value to human society has to be addressed. These questions must all be thoroughly considered before research is carried out and most developed countries have regulatory bodies that ensure that this happens.

It has been argued by some geneticists that because GMOs have extreme pheno-types, this is likely to make them less fit than wild individuals. The theory is that the added gene, although improving some market-relevant trait, acts like a deleterious allele at a locus and would be driven to extinction by natural selection.

References

Chen, T.T., Zhu, Z., Lin, C.M., Gonzalez-Villasenor, L.I., Dunham, R. and Powers, D.A. (1989) Fish genetic engineering: a novel approach in aquaculture. In: *Aquaculture '89 Proceedings.* National Shellfish Association, Los Angeles, California

Dunham, R.A. (1990b) Genetic engineering in aquaculture. *AgBiotech News and Information* 2, 401–406.

Fletcher, G. and Davies, P.L. (1991) Transgenic fish for aquaculture. *Genetic Engineering* 13, 331–369

Klinger, T. (1998) Biosafety assessment of genetically engineered organisms in the environment. *Trends in Ecology and Evolution* 13, 5–6.

Tiews, K. (ed.) *Selection, Hybridization and Genetic Engineering in Aquaculture,* Vol. 2. Heeneman Verlagsgesellschaft, Berlin, pp. 239–253

7

Genetic Markers in Aquaculture

The use of molecular genetic markers to address questions related to aquaculture management has found a steadily widening application in the last two decades. These markers can provide valuable information for various aspects of aquaculture practice, such as genetic identification and discrimination of aquaculture stocks, monitoring the consequences of founding and propagation of aquacultured stocks, assisting selective breeding programmes, and assessing chromosomal and gene manipulations such as induction of polyploidy and gynogenesis. Another important application of genetic markers in aquaculture is the assessment of the impact on natural populations of escaped or released cultured fish.

Markers are necessary to study genomes, conduct gene-linkage mapping, locate genes on chromosomes, isolate genes, determine gene expression, study the biochemical and molecular mechanisms of performance, conduct population genetic analysis and apply marker-assisted selection. Knowledge of gene locations can be utilized along with physical mapping to clone useful genes by positional cloning. However, before positional cloning of useful genes is possible, thousands of molecular markers must be identified for any aquatic species of interest.

Technology has advanced to the point that a tremendous array of biochemical and molecular markers are available to study the genetics of fish and aquatic invertebrates. One of the earliest and most tedious analyses was blood typing, but now this technique is seldom utilized. Traditional markers included isozymes, restriction fragment length polymorphism (RFLP) markers and mitochodrial DNA (mtDNA) analysis. Several powerful new types of markers have been developed, including random amplification of polymorphic DNA (RAPD), microsatellites or simple sequence repeats (SSRs), amplified fragment length polymorphism (AFLP), expressed sequence tags (ESTs) and single nucleotide polymorphism (SNP). These new DNA technologies have allowed the construction of gene maps in a matter of months, rather than the years that were the case for the construction of gene maps with conventional molecular markers, such as RFLP.

Isozymes and Enzymes

Isozymes are multiple molecular forms of individual enzymes. These multiple forms can be alleles of one another at a single locus—allozymes—or can be products of different loci where there are multiple copies of genes making the same enzyme or enzyme sub-units. Temporal differences in isozyme expression exist, which can be utilized in the study of developmental genetics—spatial or tissue-specific expression as well as allelic variation. Isozyme and enzyme analyses are technically easy, but are limited in both the numbers of loci available and polymorphism. For example, isozyme variation is low in Nile tilapia.

However, one major advantage is that genetic variation is being measured, which is directly related to protein products that actually affect performance. For example, Hallerman *et al.* demonstrated that isozyme variation is associated with growth rate in channel catfish. Isozyme variation has also been linked with disease resistance, temperature tolerance, developmental speed and salinity tolerance in fish.

Additionally, when 1+-year-old smolts of five hatchery strains of Atlantic salmon were released into a Danish river in 1996, three of the strains went to sea almost immediately, but two strains waited for more than 2 weeks before migrating. Differences in the temporal expression of gill enzyme development were highly correlated with migration pattern, and early-migrating strains reached high enzyme activity earlier than late-migrating strains. The strains with delayed enzyme development and migration exhibited a delayed regression of seawater tolerance compared with the early strains.

Additionally, northern and southern populations of the minnow, *Fundulus heteroclitus,* have different levels of expression of the lactate dehydrogenase (LDH)-B gene, *Ldh-B*. The northern strains, such as Newfoundland fish, have superior expression at lower temperatures, while the southern strains, such as Florida fish, have superior expression at higher temperatures. Deletion studies have been carried out to identify the approximate location within the regulatory sequence where the adaptive changes in the transcript occurred. A difference of only 1 base pair in the regulatory sequence accounted for the adaptive difference in *Ldh-B* expression between the northern and southern populations.

Another major advantage is that isozymes are inherited in a codominant fashion. This makes heterozygotes and homozygotes readily distinguishable, thus strengthening applications for gene mapping, population-genetics studies and determining parentage.

Isozymes can be separated in an electric field passed through a matrix, such as starch, cellulose acetate or polyacrylamide, based on their size, shape and charge, since most frequently different isozyme forms of the same enzyme vary in one or more of these parameters. Slices of the matrix are incubated in a specific histochemical stain to visualize the desired enzyme. Most staining procedures result in the deposition of dye at the site of enzyme activity, but a few stains involve a reverse process in which only the site of activity remains unstained, such as is the case for superoxide dismutase (SOD). Upon termination of the staining, the intensity of zones of staining reflects the proportions in which the gene products are present, provided that the staining is terminated before overexposure occurs. The resulting zymogram is genetically interpreted. Most gene products migrate towards the anode, and varying the pH of the buffer can affect the mobility of gene products and, in rare cases, the direction of migration.

Various enzymes/isozymes can be encoded by single- or multiple-locus systems. In all single-locus systems, the homozygous genotype yields a single zone of activity (band) on a starch-gel slice. Heterozygous genotypes yield multiple bands. The phenotype of heterozygotes will include the superimposition of the homozygous patterns in a diallelic situation, plus additional zones attributed to allozyme and isozyme heteromers of multimeric enzymes.

Enzymes are made of single or multiple protein subunits—series of polypeptide chains. Enzymes can be monomers, dimers, trimers, tetramers, hexamers and octamers made of one, two, three, four, six and eight subunits, respectively. Allozymes encode a single subunit, and these subunits bind in the cell or tissue to form the entire protein/enzyme. The dimeric system is the most common. The trimeric, hexameric and octameric systems are rare. Of 100 enzymes frequently examined in human genetics studies, there were 28% monomers, 43% dimers, 4% trimers, 24% tetramers and 1% octamers.

In the case of single-locus monomeric systems, the homozygous genotype, both alleles make the same subunit, resulting in a single isozyme/band. When these loci produce different allelic products in the heterozygous condition, each allele produces a different subunit, but the subunit represents the entire isozyme so there is expression of two products in the heterozygote at a 1:1 ratio.

Trimeric proteins are rare, and the trimeric enzyme system most utilized for fish is purine-nucleoside phosphorylase, which is usually a multilocus system. A heterozygote for a single-locus trimeric system would produce a 1:3:3:1 banding ratio, with the heterotrimers being more abundant than the two homotrimers.

Of course, in the case of a single-locus tetrameric system, the homozygotes will express a single band on starch gels because the gene products or subunits that combine to form the whole protein/enzyme are identical entities. In the case of the heterozygote, five subunit combinations are expected in a ratio of 1:4:6:4:1 reflecting random combinations of AAAA, AAAB = AABA = ABAA = BAAA, AABB = ABAB = BAAB = BABA =

BBAA = ABBA, ABBB = BABB = BBAB = BBBA, and BBBB. The three heterotetramers should have elec-trophoretic mobilities spacing them equiva-lently between the homotetramers. If limited separation of the homotetramers occurs, the heterotetramers may not be discrete and will appear as an elongated smear of activity on the gel slice.

In the case of single-locus dimeric systems, the homozygotes express a single band on starch gels because the gene products—subunits that combine to form the whole protein/enzyme—are identical. The resulting bands are homodimers. Heterozygotes express two allelic subunits, and these sub-units generated by the homologous chromosomes usually bind together in random fashion to produce the enzyme proper. Heterozygotes express both parental homodimers and a het-erodimer from the two allelic subunits in a 1:2:1 ratio, with the heterodimer being the most intense band. The heterodimer should be intermediate to the two homodimers in electrophoretic mobility.

Multilocus systems are more complex. The gene products or subunits from different loci may combine if they are expressed in the same tissue and cell type, thereby forming homomers within loci, heteromers within loci and heteromers between loci. Alternatively, expression can be strictly tissue-specific, or expression can be temporally or spatially separated within a tissue, preventing formation of heteromers between loci. Since they are under different regulatory control, the quantity of product produced by each nuclear locus may not be equivalent, as is the case for heterozygotes at a single locus. Thus, ratios of banding intensity in multilocus systems can be, but are not necessarily, a predictable symmetrical series. Interactions of products in a multilocus system will not usually obscure the predictable ratios of activity of products within a given heterozygous locus, except in unusual cases where duplicate loci have allelic variation in the same tissue and when distinct loci have alleles with the same mobility being expressed in the same tissue.

Mitochondrial and supernatant cytoplasmic loci are assembled independently; therefore there is no possibility of heteromer formation of multimeric enzymes between these two types of loci. However, some subunits of mitochondrial enzymes, such as nicotinamide adenine din-ucleotide (NADH) dehydrogenase and cytochrome C oxidase, are coded in the nucleus and transported to the mitochondria after transcription and translation.

Isozymes are a strong tool for studying evolutionary genetics. Knowledge of evolutionary genetics is necessary in some cases to fully understand and interpret isozyme data.

Genes may be duplicated via the tandem duplication of a single gene or a set of neighbouring genes. Tandem duplications are rare; until they have diverged enough to express different allelic products, tandem duplications are difficult to detect.

Alternatively, genes can be duplicated and speciation will occur following polyploidiza-tion events of ancestral fish, and this has occurred in lineages of at least six orders. Species within the families Salmonidae and Catostomidae are derived from tetraploid ancestors. Salmonids are presumed to be of autotetraploid—intraspecific polyploid event—origin and catostomids of allotetraploid—interspecific hybridization and polyploid event—origin. As is the case with tandem duplicates, the salmonid genes duplicated by the polyploid event were initially identical and under the same regulatory control. In cases where these conditions have been maintained, equivalent gene products are produced by isoloci, making genetic interpretations of salmonid zymograms more complex due to this form of genetic control.

Duplicated loci can eventually diverge, both in their allelic composition and in regulatory aspects of their tissue expression, which would result in distinct loci with tissue-specific expression. As long as at least one of these loci maintains its original metabolic function, the other is free to evolve and acquire new functions through mutation and be selectively maintained. This results in the evolution of multilocus systems. Another possibility is the silencing of a duplicated locus via mutation or functional diploidization, if no selective advantage is afforded the diverged duplicate locus. Silenced loci might be able to retetraploidize and resume their former function if a favourable regulatory mutation occurs, and this may have occurred for glucose-phosphate isomerase expression in *Moxostoma lachneri.*

Another phenomenon that can complicate the interpretation of zymograms is null alleles that do not produce or encode protein products (subunits) or produce very reduced amounts of the subunits, which can yield skewed ratios of activity or unexpected proportions of certain genotypes. Of course, null alleles act as recessive alleles as they cannot be detected in the heterozygote individuals. Progeny testing is one of the best ways of detecting null alleles. Null alleles are rare but have been detected in carp, rainbow trout and oysters.

Heteropolymer restriction can also complicate interpretation of zymograms. In some cases, random association of subunits of mul-timeric enzymes does not occur and formation of heteromers is restricted. There is bias in the assembly of the multimer, with similar allelic products being more likely to combine. For example, the creatine kinase (CK) product predominating in skeletal muscle (*Ck-A* locus) often has restriction of intralocus heterodimer assembly in fishes. The heterodimeric combination is not formed, yielding heterozygotes exhibiting two zones of activity or bands—the two CK parental forms—on starch gels, as would be expected for heterozygotes for monomeric product.

The restriction of heteromer assembly among products of different loci is more common. Interlocus restriction of assembly occurs in the LDH system. In most fishes, both *Ldh-A* and *Ldh-B* products are usually present in skeletal muscle and interact to form three interlocus het-erotetramers and two homotetramers in the double-homozygous genotype. However, in several fish, such as darters, Etheostomatini, the assembly is restricted to the homotetramers (AAAA and BBBB) and the symmetrical heterotetramer (AABB—in some or all of its six possible assemblies). In *Etheostoma fonticola* and *Gyrinocheilus aymonieri,* all interlocus heterotetramers are restricted and not assembled, limiting expression to the homotetramers.

Vertebrate haemoglobins can also be restricted in their assembly. Two loci (α and β) are contributing tetrameric subunits to the haemoglobin molecule, but assembly is restricted to form only the symmetrical heterotetramer, α, β.

Restriction Fragment Length Polymorphism

RFLP was once widely used and is still very useful, and has been used to construct genetic maps of many species. Restriction endonu-clease enzymes are used in this method to directly cut the DNA at restriction sites. Base substitution at the restriction sites, insertions, deletions or DNA fragment rearrangements at or between the restriction sites cause the polymorphism. The resulting products are then separated on agarose gel, transferred to a membrane and hybridized with labelled probes to produce DNA fingerprints. The advantages of RFLP include codominant inheritance and easy interpretation and scoring. This technique is now less frequently used because it is time-consuming and requires tedious Southern blotting. Additionally, probe development is required for RFLP analysis, polymorphism is generally low and sequence information is needed if using polymerase chain reaction (PCR). This technique is too slow and tedious to generate large numbers of markers.

Mitochondrial DNA

The analysis of mtDNA variation is an alternative for studying population genetics in fish. For species such as striped bass where isozyme variation was minimal, significant mtDNA variation was observed. The mutation rate of mtDNA is about an order of magnitude higher than that of the nuclear genome, and the control region is particularly hypervariable, thus allowing studies on recent evolution.

Since the mitochondrion is the major site of cellular respiratory metabolism and a possible source of maternal effect, genetic improvement programmes should be concerned with mtDNA as well as nuclear DNA. MtDNA analysis often revealed genetic differences among populations of fish that were homogeneous for isozyme variability. Three types of polymorphisms can be detected for mtDNA in fish: length polymorphisms, restriction-site polymorphisms caused by base-pair additions, deletions or both, and heteroplasmy.

MtDNA heteroplasmy is the existence of more than one form (genotype or haplotype) of mtDNA in an individual. Natural mtDNA heteroplasmy has been observed in bowfin, *Amia calva,* American shad, *Alosa sapidissima,* striped bass, white sturgeon, *Acipenser transmontanus,* dwarf cisco, *Coregonus artedii,* red drum, *Sciaenops ocellatus,* Atlantic cod, *Gadus morhua,* and anchovy, *Engraulis encrasicolus.*

Heteroplasmy could result from one of two mechanisms. On rare occasions, mtDNA can be inserted and inherited from the male parent—paternal leakage. A second mechanism would be a mutation in the mtDNA genome, with some type of selective force or random process resulting in an increased population of the mutated mtDNA until it was detectable. It is likely that heteroplasmy is more frequent than what is detected because the secondary haplotype could be at frequencies too low to detect.

MtDNA does encode genes and could affect the performance of aquatic organisms. Sequence and restriction analysis readily detects variation in mitochondrial genes. Genetic variation exists for the *ND5/6* gene of mtDNA in different strains of *Oreochromis niloticus.* The indices of haplotype diversity and nucleotide diversity of *O. niloticus* were 0.69 ± 0.10 and 0.03 ± 0.10, respectively.

Randomly Amplified Polymorphic DNA

RAPD markers are polymorphic DNA sequences separated by gel electrophoresis after PCR, using one or a pair of short (8-12 base pairs (bp)) random oligonucleotide primers. Polymorphisms are a result of base changes in the primer-binding sites or of sequence-length changes caused by insertions, deletions or rearrangements. RAPD is very powerful in detecting large numbers of polymorphisms because oligonucleotide primers scan the whole genome for perfect and subperfect binding sites in a PCR reaction. When two binding sites are close enough (3000 bp or less), a RAPD band is produced on the gel. Each RAPD primer usually amplifies several bands, some of which are polymorphic in even closely related populations, which can be either tremendously advantageous or disadvantageous.

RAPD markers are expressed and scored as dominant alleles. The amplified DNA product is scored based on size and presence. A polymorphism occurs when a band is present in one parental type but absent in the other. Even if a homologous fragment exists in the other parent, exhibited as a band with a different size, it would be scored as a distinct marker, although it actually represents the same locus or the same general location of the DNA sequence. Technically, RAPDs are not genes or alleles as they do not code for gene products. A potential disadvantage for RAPD analysis is that these dominant banding patterns fail to distinguish between heterozygous and homozygous individuals. Of course, inheritance of the markers could be verified by progeny testing, but this is not simple because of the large number of bands. Potentially, sequence-tagged site (STS) markers could be developed from the RAPD markers by cloning and sequencing of the RAPD markers, and the STS markers would be codominant, increasing the power of the analysis.

RAPD markers are particularly useful for efficient, economic, non-radioactive DNA fingerprinting of genotypes for the determination of genetic relationships and rapid construction of genetic linkage maps. RAPD does not require any known probes or sequence information necessary for RFLP or microsatellite analysis. RAPDs are highly polymorphic and the technique is simple and fast. RAPD markers meet the requirements of a good marker system: the generation of large numbers of polymorphic markers, simplicity, economical, reproducible and normal Mendelian inheritance.

The primary drawback with RAPD analysis is the potential for reduced reproducibility because of the use of short random primers (usually ten nucleotides long), which necessitate lower annealing temperatures for PCR. Consequently, such short primers can bind to both their perfectly homologous binding sites and sites that are not completely homologous—suboptimal (nonspecific) regions—especially during the first few cycles. This creates the risk of nonspecific amplification and possible variation in results among experiments between different aliquots of the same sample at different dilution, and especially among laboratories. However, reproducibility was excellent for RAPD analysis of ictalurid catfish within the size range of 400-1500 bp. Amplified products larger than 2 kilobase pairs (kbp) and smaller than 200 bp showed lower reproducibility. The bands were reproducible in number and over time. Higher concentrations of DNA template and primers led to amplification of more bands, making scoring more difficult. However, quantifying DNA before RAPD analysis and using the same concentration of primers gave consistent and reproducible results. The concentration and purity of genomic DNA templates used for PCR are the major factors for obtaining reproducible results. Genomic DNA should be prepared using a constant procedure, and the quantity of DNA should be determined before starting RAPD. Primer concentration should be kept constant to obtain reproducible results. Another weakness is that the testing of large numbers of primers is required to generate large numbers of markers.

RAPD has been used in the guppy *Poecilia reticulata,* tiger barb, *Barbus tetrazona,* and medaka, *Oryzias latipes,* to study genetic variation. Only low levels of RAPD polymorphism existed among strains of channel catfish and strains of blue catfish. Fewer than 5-10% of the bands appeared to be polymorphic in some strains within each species. Fixed differences in RAPD genotypes were found for hellbender (salamander) and for striped bass (R.A. Dunham, H. Kucuktas and Z. Liu, unpublished). It appears that the likelihood of finding strain-specific markers is greater via RAPD analysis than via isozyme analysis.

As expected as it is for all types of genetic markers, interspecific RAPD variation was much greater than intraspecific RAPD variation for channel catfish and blue catfish. More than 40% of the bands were polymorphic when comparing the two species. Each random RAPD primer amplified 5.3 bands from channel catfish and 5.5 bands from blue catfish. About 47.3% of the amplified bands from channel catfish and blue catfish with random RAPD primers were species specific.

Liu *et al.* examined the inheritance of RAPD sequences in channel catfish-blue catfish hybrids. All polymorphic paternal and maternal bands were amplified, and the sum of all RAPD bands from both parents existed in RAPD profiles of F_1 hybrids, indicating full penetrance, the dominant nature of RAPD markers and a dominant Mendelian pattern of inheritance in the F_1 progeny. All polymorphic paternal and maternal bands were amplified. In all cases, channel catfish RAPDs segregated in F_2 and reciprocal F1 X blue backcrosses and blue catfish RAPDs segregated in F_2 and reciprocal F_1 X channel backcrosses in expected ratios, confirming the dominant nature of the RAPD markers and the Mendelian inheritance.

Reciprocal F1 hybrids (channel catfish female X blue catfish male and blue catfish female X channel catfish male) were also evaluated because of paternal predominance being prevalent for many of the phenotypic traits of the reciprocal hybrids. However, all RAPD bands were transmitted into the F_1 hybrids, regardless of the sex of the parents, and there was no apparent relationship between RAPD markers and paternal predominance in channel-blue hybrids.

By adjusting experimental conditions, almost any RAPD primer can be made to yield data. However, certain analyses, such as gene mapping, require testing of large numbers of individuals for each primer, and it is impractical and too time consuming to elucidate and implement all of these conditions or to utilize low-yielding primers. RAPD primers for gene-mapping analysis need to fulfil certain stringent criteria. A RAPD primer for gene mapping should generate a reasonable number of RAPD bands, detect high levels of polymorphism, be easy to use without special conditions and have high reproducibility. The number of bands amplified is important because, if too few bands are generated, a large number of primers and gel runs would be required to produce enough data points. If too many bands are amplified, it may make scoring and analysis difficult or impossible. High levels of polymorphism can reduce the numbers of primers and gel runs required to give the same numbers of markers.

RAPD markers must also pass some stringent tests to be useful for gene-mapping analysis. A useful RAPD marker for gene mapping needs to be polymorphic and seg-regational, highly reproducible, prominent in band intensity and well separated from the other bands for ease of scoring. Liu *et al.* found that 52% of the RAPD primers tested for the blue catfish-channel catfish system were good to excellent primers, generating seven or eight RAPD markers per primer. In contrast, RAPD was inefficient in producing intraspecific polymorphic markers in catfish and would not be a good choice of molecular marker for intraspecific gene-mapping strategies.

A RAPD primer's usefulness for application across multiple species may be dictated by genetic distance. The primers evaluated by Liu *et al.* worked well for both channel catfish and blue catfish, two highly related species. However, selected primers that were used for gene mapping in zebra fish performed worse than random primers in the channel catfish-blue catfish evaluation, indicating that primers appropriate for one family of fish are not the most appropriate for another family. Similarly, high-GC primers were less useful than random RAPD primers in ictalurid catfish.

Amplified Fragment Length Polymorphism

AFLP combines the strengths of RFLP and RAPD. Genomic DNA is digested with two restriction enzymes *EcoRI* and *MseI,* suitable adaptors are ligated to the fragments and the ligated DNA fragments are selectively amplified with different primer combinations; then the products are resolved by gel electrophoresis. AFLPs are highly polymorphic and the technique is simple and fast. The molecular bases of AFLP polymorphism are base substitutions at the restriction sites, insertion or deletion between the two restriction sites, base substitution at the pre-selection and selection bases and chromosomal rearrangements. The advantages of AFLP include its PCR-based approach, requiring a small amount of DNA; no requirement for any known sequence information or probes; and the specific amplification of a subpopulation of the restriction fragments because the long PCR primers in the procedure allow high annealing temperature and high repeatability. Perhaps the greatest advantage of AFLP analysis is that it is capable of producing large numbers of polymorphic bands in a single analysis at a relatively low cost per marker. The generation of hundreds up to thousands of bands with limited numbers of primer combinations makes AFLP a very efficient and economical system for genetic analysis. There can be more than 100 loci per primer combination and over 4000 primer combinations to evaluate, and more than one restriction enzyme can be used, making it possible to generate tremendous numbers of markers. AFLP bands are widespread and evenly distributed, giving near genome-wide coverage. AFLP is highly reliable because it combines the advantage of RFLP and RAPD and is devoid of the disadvantages of the slow speed and low levels of polymorphism of RFLP and the low reproducibility of RAPD. The AFLP procedure allows the genetic analysis of closely related populations. The disadvantages of AFLP are that they are dominantly inherited and that more technically demanding and specialized, expensive equipment, such as DNA sequencers, is required.

In some ways, these new DNA technologies can be almost too powerful. If enough markers are utilized, probably every population or sample can eventually be distinguished from any other population or sample. Criteria need to be established to match the new technologies to define when populations are actually differentiated. This will not be an easy concept to develop.

Liu *et al.* indicate that, among the polymorphic DNA fragments, two subsets can be distinguished: the presence or absence of bands and band-intensity polymorphisms. Presence or absence polymorphisms result from the gain or loss of restriction sites, insertions, deletions or reversions between restriction sites, or from the fact that selective nucleotides of the primer used in the AFLP procedure and the sequences adjacent to the restriction site are not complementary.

The intensity of the polymorphism is very difficult to score. In reality, even if band intensities could be quantified, the intensities of the amplified band are not really polymorphisms.

If the samples are obtained and analysed in exactly the same way, the DNA at the location of a specific AFLP marker should be amplified equally among all individuals possessing that sequence, or allele. If band intensities could actually be quantified, there would be two explanations for varying band intensities. One case would be the difference between homozygous individuals, which would have two copies of the marker, and heterozygous individuals, which would have one copy of the marker. The cumulative effects of different chromosomal locations generating AFLP fragments of exactly the same size could also generate bands of different intensities. This would indeed be real genetic variation but would not be specific to a single chromosomal location, or locus. Differences in band intensity could also be the result of a combination of both these phenomena, making interpretation potentially difficult or impossible.

In the case of catfish, the polymorphic paternal and maternal bands showed dominant Mendelian pattern of inheritance in the F_1 interspecific progeny of channel and blue catfish. Few segregations of polymorphic AFLP markers were observed in F1 individuals, thus indicating that the majority of the polymorphic loci were homozygous in the parental species. The majority of the AFLP markers must represent species-specific markers. The markers segregating in F1 individuals represented allelic variation within a species. Those AFLP markers were heterozygous in the paternal or maternal parents. Since low levels of polymorphisms were detected within either channel catfish or blue catfish, most AFLP loci associated with interspecific variation were homozygous within the species. This also indicates that multiple chromosomal locations for the same-sized bands utilizing the same primers must be a rarity; otherwise the transmission of the AFLP markers to the F1 hybrids would have been more complicated. That leaves no logical genetic mechanism that would generate bands of different intensities in ictalurid catfish. This may be true for other species of fish as well. The results also indicated that chromosomes of channel catfish and blue catfish paired properly and their AFLP markers followed Mendelian inheritance.

The application of AFLP markers in genetic linkage and quantitative trait locus (QTL) mapping and analyses of genetic resources has greatly sped research in these areas. The efficient, rapid, economical development of AFLP markers linked to disease resistance genes, has allowed application of these markers in marker-assisted selection programmes.

Liu *et al.* examined the characteristics of the production of high-quality AFLP markers. AT-rich selection bases were more associated with a lower quality of AFLP fingerprinting. With the exception of two primer combinations, all of the lower-quality primer combinations were from primer combinations with primers M-CAT or M-CTT. This may indicate that the genomes of channel catfish and blue catfish are AT-rich. The AT-richness would create a greater number of amplified bands when the selective bases are AT-rich, especially at the most 3' position. AT-rich primer combinations also resulted in greater numbers of total amplified bands in barley using AFLP. The weak intensities of these bands may be caused by low efficiency of primers with AT 3' ends in PCR reactions. However, if the AT-richness does cause amplification of a greater numbers of bands, the intensities should be similar among all amplified bands, with the exception of bands amplified from highly repetitive elements. That was not the case for channel and blue catfish. Alternatively, these weak bands could be from non-specific priming at the mismatched sites. Most of the poor primer combinations had T as the 3'-terminal base in the primer. Kwok *et al.* indicate that primers with 3'-terminal T mismatches can be efficiently utilized by *Taq* polymerase when the nucleotide concentration is high. Amplification of a T mismatched with a C, G or T may be initiated frequently, although at a lower efficiency than with a matched base. The catfish results of Liu *et al.* are consistent with this explanation. If this is correct, T should be avoided from the terminal selective bases for the design of AFLP primer kits.

The primers and selection bases can affect the variation of AFLP profiles. For channel catfish and blue catfish, *EcoRI* primers had a large effect on the total number of amplified bands—49-267 bands Eight *EcoRI* primers produced 69—161 total bands per primer combination when combined with the eight *MseI* primers, and the eight *MseI* primers produced 93–145 total amplified bands when combined with the eight *EcoRI* primers.

The terminal selective bases have large effects on both the total numbers of amplified bands and their reproducibility. Terminal T exhibited the lowest levels of selectivity, producing the largest numbers of amplified bands and the highest levels of background bands in ictalurid catfish. Among eight *EcoRI* primers, E-ACG produced the least numbers of total amplified bands. Eight *MseI* primers produced a mean of 69 bands per primer combination. E-ACT produced the largest total number of amplified bands, with a mean of 161 bands per primer combination, when used with the eight *MseI* primers. Among eight *MseI* primers, M-CTC generated the lowest numbers of amplified bands, with a mean of 93 bands per primer combination, and M-CTT produced the largest numbers of total amplified bands, with a mean of 145 bands per primer combination. The terminal-T selection base generated large numbers of amplified bands compared with other selective bases.

The position of selective bases, rather than the AT-richness, has the greatest effect on AFLP fingerprinting patterns. Although the percentage of G/C and A/T bases in the selection bases of *EcoRI* primers used by Liu *et al.* when analysing channel catfish and blue catfish was the same (50%) as in the *MseI* primers (50%), their positions are different. Two of the eight *EcoRI* primers tested harboured A/T at their 3' terminus, while four of the eight *MseI* primers had a terminus of A/T bases. Similarly, at the second-most 3' position, *Eco*RI primers had two A/T bases out of eight. In contrast, *MseI* primers had eight A/T bases out of eight. Combining the second and the third most terminal bases, *EcoRI* primers had four of 16 being A/T bases and *MseI* primers had 12 of 16 being A/T bases. This may explain the larger variation of total numbers of amplified bands generated from *EcoRI* primers (variation from high G/C to high A/T at the terminal bases) than that from *Mse* I primers (variation from medium A/T to high A/T at the terminal bases).

The polymorphic rates of the AFLP bands were inversely correlated to the total numbers of amplified bands in blue and channel catfish. Primer combinations that produced large numbers of amplified bands had lower percentages of the bands being polymorphic. Primer combinations that produced small numbers of amplified bands generated higher rates of polymorphism. Eight *EcoRI* primers produced a mean of 33-62% polymorphic bands. The eight *MseI* primers resulted in a mean of 30-50% polymorphic bands. Among the *EcoRI* primer combinations, primer combinations that produced the smallest numbers of total amplified bands (X = 69) had the highest mean rate of polymorphism—62%. Similarly, among the *MseI* primer combinations, primer combinations that produced the smallest numbers of total bands (X = 96) exhibited the highest mean polymorphism -49%. In contrast, the primer combinations that produced the largest mean number of amplified bands—161—among the *EcoRI* primer combinations had the lowest mean polymorphic rate (33%) among the *Eco* RI primer combinations. Similarly, the primer combinations with *MseI* primer 8 produced the highest mean total of amplified bands -145 per primer combination—but their mean polymorphic rate (30%) was also the lowest among the *MseI* primer combinations. Again, the total numbers of amplified bands appeared to be related to the terminal selective bases. Primers with the terminal T produced large numbers of amplified bands, which was correlated with a lower percentage of polymorphic bands than primers with G, C or A terminal bases.

High reproducibility is required for good DNA markers. Liu *et al.* tested the reproducibility of AFLP markers in channel and blue catfish by using DNA templates from different individual fish isolated at different times. High levels of reproducibility were observed as all individuals tested over time always exhibited identical banding patterns.

Different numbers of bands may be observed as a function of different radioisotopes. Generally, the same AFLP profiles were obtained with either ^{32}P or ^{33}P for ictalurid catfishes. However, ^{33}P allowed detection of more AFLP bands than ^{32}P This difference is probably related to the low-energy radioisotope ^{33}P allowing longer exposures than the high-energy

^{32}P For instance, with a 1-week exposure of ^{33}P it is possible to obtain reasonably clean autoradio-grams, while a 1-week exposure with ^{32}P generated overexposed dark autoradio-grams. Longer exposure allowed some weaker bands to be detected, which otherwise were not visible or were just too weak. While the detection of more bands assisted robust analysis of many loci simultaneously, too many bands could make analysis extremely difficult, especially for adjacent markers all segregating. Another factor for the reproducibility of AFLP bands is the size of amplified products. Generally, amplified products with large sizes displayed on the top of sequencing gels have lower reproducibility. Often the large fragments were not efficiently amplified to generate strong bands. AFLP bands of 50-500 bp exhibited the highest reproducibility. Large variation in the number of bands amplified was observed when different primer combinations were used. This is not a reproducibility issue, but is relevant for primer selection for genetic linkage analysis.

Similarly to RAPD analysis, intraspecific poymorphism for AFLP can be relatively low. Intraspecific polymorphisms in ictalurid catfishes were generally less than 10% of all bands. However, over 50% of AFLP bands were polymorphic between channel catfish and blue catfish, and were suitable for use in mapping analysis using the interspecific hybrid system.

As expected and similarly to the case with RAPD markers, there was no relationship between AFLP genetic variation and paternal predominance in channel-blue catfish hybrids. Even though strong paternal predominance was observed for the morphological and meristic traits of interspecific F_1 hybrids, transmission of AFLP markers to F_1 was normal, and the two reciprocal F_1 hybrids of channel catfish X blue catfish had identical AFLP profiles. Most AFLP markers were inherited in expected Mendelian ratios in F_2 or backcross hybrids.

AFLP can generate large numbers of DNA bands without any previous knowledge of a fish genome. This technology offers a robust analysis of genomes, allowing approximately 100 catfish genomic fragments to be displayed in a single analysis. The high levels of interspecific polymorphism between the channel catfish and blue catfish suggest that AFLP analysis should be highly useful for generating a large number of markers for genetic linkage analysis, using the interspecific hybrid system to produce mapping populations. Exactly as in the case for RAPD molecular markers, the low level of intraspecific polymorphism makes AFLP an inefficient marker system for mapping analysis using an intraspecific mating design. More intraspecific polymorphism was detected with AFLP analysis than with RAPD analysis for channel catfish and blue catfish.

Segregation of channel catfish and blue catfish AFLP markers followed expected ratios except that there was strong segregation distortion for two blue catfish markers in the F_2 and backcross hybrids selected for increased growth rates. These two markers were either not detected or were detected at much lower frequencies than expected. Since the F_2 and backcross hybrids were selected for increased growth rate, these markers and the corresponding markers in channel catfish may be linked to growth loci. The segregation could be caused by selection pressure for increased growth or possibly by natural selection. If there were genes or genomic sequences in blue catfish that had a negative effect on growth rate in comparison with the channel catfish sequences, selection for increased growth rate should select against these genes or genomic sequences. Since both markers were absent from the individuals analysed, the two markers may be linked or they could be independently linked to two separate growth-encoding loci. Alternatively, the segregation distortion may be due to natural selection for survival. If some genes or genomic sequences have a negative effect on survival, they would be selected against and detected at a reduced frequency.

Microsatellites

Microsatellites are simple-sequence, tandem repeats of 1-6 bp. The molecular basis of microsatellites is the difference in number of repeats. Microsatellite markers are ideal molecular markers because they are highly polymorphic, evenly distributed in genomes and codominantly inherited. They are highly useful among various types of DNA markers because their high rate of polymorphism and codominant inheritance allow precise genetic analyses, increase mapping accuracy, maximize the genetic information generated and allow lineages of individuals or families (individual spawns) to be accurately traced. High levels of polymorphism also indicate that microsatellite markers may be highly useful for population genetic analysis and strain identification. Microsatellite loci are short in size, facilitating genotyping via PCR. Their disadvantage is that microsatellite analysis requires great effort, time and expense in library construction, screening, sequencing and PCR primer analysis, and they may have nonspecific bands. Characterization of large numbers of microsatellites for the construction of genetic maps with high resolution is a tedious and strenuous task.

The majority of microsatellite loci can be amplified in both channel catfish and blue catfish, suggesting evolutionary conservation between these two catfish species, and microsatellite markers can be used as codominant markers in the interspecific hybrid system for genetic linkage analysis. Liu *et al.* confirmed that microsatellite markers are inherited as codominant markers in catfish. Microsatellites were highly polymorphic in both channel catfish and blue catfish. The microsatellites were developed from channel catfish. Only 10% of microsatellite loci could not be amplified in blue catfish. In the case of these loci, the divergence may be quite dramatic so that primers could not bind at all, or it could be minor so that primer binding might require lower temperatures. One microsatellite locus, Ip351, was amplified at 40°C to produce allelic fragments, which were not amplified from blue catfish at 50°C or higher temperatures, indicating conservation of the locus with minor base substitutions at the primer-binding sites.If large percentages of primers can amplify across species borders, high levels of genomic conservation are indicated. The high levels of genomic conservation of channel catfish and blue catfish were consistent with the indistinguishable karyotypes of the two species.

Conservation of microsatellite loci among closely related species is expected, although successful amplification of these loci across species boundaries depends on the conservation of primer sequences. In some genera, families, orders or phyla, microsatellites are deficient and primers from one species will often not work in related species. Sequence variations at the primer region can significantly affect the success of PCR amplification. Base changes at the 3' end of the primer-binding sites are more critical than at the 5' end. Caution should be exercised when interpreting the existence of null alleles, because they can be obtained from amplifications from only one allele when one or both primers fail to bind to the alternative allele, thus skewing the data towards a higher frequency of homozygosity.

Genomic conservation of microsatellite loci has also been compared among channel catfish, blue catfish, white catfish *(Ameiurus catus)* and flathead catfish *(Pylodictus olivaris)*, all in the family Ictaluridae. The microsatellite loci were highly conserved in all genera tested from Ictaluridae. All channel catfish primers tested successfully amplified genomic DNA from flathead catfish, and 86% of the channel catfish primers successfully amplified the genomic DNA from white catfish. Southern blot analysis confirmed allelic amplification. If the amplification is allelic, all the amplified bands from all species should harbour microsatellite sequences and thus should hybridize to the microsatellite probes. All amplified bands from channel catfish, flat-head catfish and white catfish hybridized to the $(CA)_{15}$ probe, confirming allele-specific amplifications.

Ictalurid catfish have similar genomic organization. Microsatellites appear to evolve free of selection pressure, and thus evolve towards high evolutionary divergence because they are non-coding sequences. This may be the general case; however, some microsatellites are found within coding regions and in this case might be subject to strong selective pressures. The recent discovery of high conservation of microsatellite marker loci across wide taxonomic borders in fish may indicate either that such sequences evolve under unknown selection pressure or that sequence evolution in aquatic animals is slower than in land animals.

Various classes of microsatellites exist and can be found at varying frequencies. In eastern oysters, dinucleotide motifs within microsatellites were dominated by AG, trinucleotide microsatellites had all possible motifs in equal frequencies and tetranucleotides were more prevalent than trinucleotides and were strongly associated with specific repetitive sequences, which was not the case for other classes of microsatellites.

Expressed Sequence Tags

ESTs are short, single-pass complementary DNA (cDNA) sequences reverse-transcribed from mRNAs and generated from randomly selected cDNA-library clones. The single-pass sequencing is at both the upstream and the downstream segments of cDNAs. The basis of EST analysis is that specific mRNAs and transcript quantities vary in different tissues, in different developmental stages or when the organism faces different environmental conditions. Characterization of ESTs is a relatively easy and rapid way for identification of new genes in various organisms. Because of the relative ease of EST analysis, the EST database is the fastest-growing division of GenBank. More than 415,000 human ESTs have been characterized. Extensive EST analysis is not only an efficient way to identify genes, but is also powerful for the analysis of their expression. ESTs indicate when, where and how strongly genes are expressed, and each EST represents a gene, so they can be used in functional genomics analysis. ESTs are particularly useful for the development of cDNA microarrays, which allow the study of differentially expressed genes in a systematic way. EST analysis is useful for comparative genomics by determining orthologous counterparts of genes through evolution. Profiling of expression provides a rapid means of examining gene expression or differential gene expression in specific tissue types, in biological pathways, under specific physiological conditions, during specific developmental stages or in response to various environmental challenges. ESTs are also efficient molecular markers for genomic mapping, and microsatellites can be found within ESTs, making ESTs even more useful for gene-mapping research. However, only 4.6% of ESTs characterized from the skin of channel catfish contained microsatellites. Disadvantages of EST analysis are that a large amount of preparatory work is required, the cDNA library may not contain transcripts of low abundance and normalization of the cDNA library may be necessary.

EST analysis greatly speeds the pace of genomics research. Implementing a transcriptomic approach, all of the 47 ribosomal protein genes in the 60S ribosome and all of the 32 ribosomal protein genes in the 40S ribosome of channel catfish were characterized in a few months, which would have taken years of analysis, requiring much greater resources to accomplish, prior to the development of EST procedures. Such systematic EST analyses allowed identification of alternative spliced transcripts as well as alternatively polyadenylated transcripts, demonstrating the greater power and value of EST analysis than are given by only surveys of genes and their expression.

Microsatellites have been discovered in ESTs generated from a variety of tissues in channel catfish. The percentage of ribosomal protein genes (ESTs) containing microsatellites was 4.6%. Ribosomal protein S16 of channel catfish contained a compound microsatellite composed of CT and CA repeats. The microsatellite locus for this protein had both types of simple sequence repeats. They were short and both were expanding. The CT repeat had genotypes ranging from eight to 11 repeats, while the CA repeat was found to have genotypes with four to five repeats. This polymorphic microsatellite should allow the genomic mapping of the S16 gene.

Microsatellites of channel catfish are also found in muscle-specific genes. The catfish myostatin gene possesses multiple microsatellites, making this gene highly polymorphic and potentially useful for gene mapping and probably marker-assisted selection. These microsatellites were found in numerous regions of the gene, including the upstream untranslated region (UTR), exons and introns. The upstream untranslated region contained highly repetitive sequences, including several hundred base pairs of simple sequence repeats with a simplified consensus of TGGTAG. One (CAG) repeat was found within the first exon encoding a polyglutamine tract. Three microsatellite sequences existed in the first intron with $(GTTT)_7$, $(TG)_{11}$ and $(TA)_{38}$ repeats. One microsatellite repeat of $(GAA)_7$ existed in the second intron. In the 3'-UTR, one (AC) repeat was present.

Single Nucleotide Polymorphisms

SNP is caused by base variation among individuals at any site of the genome. This single base variation can be determined by DNA sequencing, primer extension typing, the designing of allele-specific oligo and gene-chip technology. The advantages of SNP are that SNP sites are abundant throughout the entire genome (3×10^7 different sites have SNP in humans), they are highly polymorphic and the SNP analysis is the only system that identifies every single difference or polymorphism among individuals. SNP analysis has several disadvantages, including the need for sequence information, the necessity of probes and hybridization, high expense and difficult genotyping.

The channel catfish myostatin gene not only contained microsatellites, but also had many SNPs. Many of these SNPs were neutral and did not change amino acid sequences, but five SNPs caused changes in amino acid sequences.

Relative Costs of Different Markers

In some cases, costs may dictate what type of markers may be used for a genetic application. RFLP, RAPD, AFLP, microsatellites and SNP were compared for their relative costs. The primary advantage of RFLP and RAPD is the very low start-up costs. When considering the start-up costs, the analysis using RFLP and RAPD is going to be much less expensive until you reach about 10,000 samples for 50 markers. Once start-up costs are met, if a small number of individuals—ten to 100—are compared for anywhere from one to 50 markers, the costs are similar for all five techniques. RFLP and RAPD costs are about 25% less for this number of samples compared with the other three technologies; however, the quality and power of the data sets are lower for RAPD and RFLP. Increasing the sample size from ten to 100 individuals does not have a large impact on costs. Increasing the sample size from 100 to 1000 about doubles the costs. At 1000 individuals, the number of markers starts to have an impact on cost. Increasing the number of markers from ten to 50 increases costs by about 25% for RFLP, RAPD and microsatellites, but does not have much impact on AFLP and SNP. Costs double again when going from 1000 to 10,000 individuals, except that costs rapidly increase an additional 2.5-fold for RFLP, RAPD and microsatellites and about 50% for SNP when increasing from ten to 50 markers. Number of markers does not affect cost for AFLP because of its ability to generate many markers with limited primers.

Isozyme costs are about $16,000/160 samples for 40 loci. Therefore, for small sample numbers isozyme analysis is going to be more cost-effective than DNA markers, and the biological meaningfulness is going to be greater. However, the ability to find fixed differences between populations will be vastly inferior for isozymes compared with RAPD and AFLP and inferior to microsatellites for tracing lineages. Once sample sizes reach 1000 or more, the DNA analyses are more cost-effective than isozymes, with the same advantages and disadvantages for biological information generation as for the smaller sample sizes.

Relative Effectiveness of Markers

The various markers have different strengths and weaknesses. Yuan *et al.* compared RFLP, RAPD, SSR and AFLP for their effectiveness in delineating inbred lines of maize. The information content, as measured by expected heterozygosity and mean number of alleles polymorphic, was best for SSR analysis. AFLP resulted in the lowest polymorphic values. The genetic-similarity trees were highly correlated except for those generated by RAPD. Yuan *et al.* concluded that AFLP was the most efficient system because of the large number of bands generated and that both AFLP and SSR could replace RFLP for maize population genetic analysis.

Desvignes *et al.* examined allozyme and microsatellite gene frequencies in domestic carp strains in France and the Czech Republic. French strains had lower heterozygosity but a higher number of alleles. The two marker systems generated data giving similar results and conclusions, but the microsatellites better discriminated strains within and between countries.

Allendorf and Seeb examined 21 allozyme polymorphisms, 15 nuclear DNA polymorphisms and mtDNA variation in four Alaskan populations of *Oncorhynchus nerka*. Concordance was obtained among markers in the amount of genetic variation within and among populations, with the striking exception of one allozyme locus (sAH), which exhibited more than three times the amount of among-population differentiation as other loci. Allendorf and Seeb concluded that information should be gathered on many loci for population comparisons, but that the type of locus or marker system is of secondary importance.

Microsatellite markers underestimate genetic divergence among populations when gene flow is low. Microsatellites are extremely powerful but discriminate less well at the species level than at the subpopulation level. Balloux *et al.* examined genetic divergence among two chromosome races of the common shrew in which gene flow is reduced by a number of factors, including male hybrid sterility. The genetic divergence of these two races of shrew estimated from mtDNA, proteins and karyotypes was much larger than that estimated from microsatellites, with the exception of one microsatellite located on the male (Y) chromosome. Balloux *et al.* utilized computer simulations to show that the discrepancy arises mainly from the high mutation rate of microsatellite markers for F statistics and from deviations from a single-step mutation model for R statistics.

Gerber *et al.* compared dominant markers, such as AFLPs, with codominant multiallelic markers, such as microsatellites, for reconstructing parentage. Both sets of markers produced high exclusion probabilities; dominant markers with dominant allele frequencies in the range 0.1-0.4 were more informative. Not unexpectedly, dominant markers were less efficient than codominant markers for reconstructing parentage, but can still be used with good confidence when loci are deliberately selected according to their allele frequencies.

Controversy and uncertainty surround the issue of whether or not a population at risk of extinction is actually worth saving, given that resources are limited and priorities must be set. Genetic uniqueness and genetic diversity are often cited as the appropriate criteria for expending effort to save populations. The genetic diversity and uniqueness of neutral markers may have no correlation with the quantitative genetic variation, which is the actual basis of adaptation and short-term adaptation. McKay *et al.* found that small, peripheral populations of the rare plant sapphire rockcress are genetically adapted to local microclimates, and that local adaptation occurs despite the absence of divergence at almost all marker loci and very small effective population sizes, as indicated by extremely low levels of allozyme and DNA sequence polymorphism.

Comparison of living and fossil samples of marine snails revealed that the northern populations have come to be dominated by snails with a new, probably adaptive, thick-shelled morphology, and have become morphologically more diverse despite being relatively uniform at the mtDNA-marker loci. Similar conclusions can be drawn from data on freshwater fishes, in which post-Pleistocene colonization of new habitats has also led to evolutionary divergences.

Freeland and Boag found that differentiation of the Darwin's ground-finch species based on morphological data is not reflected in either mtDNA or nuclear DNA sequence phylogenies, and inferred genealogies based on mitochondrial and nuclear markers are not even concordant with each other. This is obviously another example illustrating that DNA markers do not always correlate with important quantitative and qualitative traits, which Lewontin predicted based on the theory that the two measures should be very poorly correlated.

Similarly, estimates of molecular and quantitative genetic variation were essentially uncorrelated in natural populations of *Daphnia*. Although molecular markers provided little information on the level of genetic variation for quantitative traits within populations, they may be valid indicators of population subdivision for such characters, as molecular measures of population subdivision gave conservatively low estimates of the degree of genetic subdivision at the level of quantitative traits. Reed and Frankham examined 71 sets of data where quantitative variation and marker variation were both estimated. The mean correlation between molecular and quantitative measures of genetic variation was weak (r = 0.22), and there was no significant correlation between molecular variation and life-history traits (r = 0.11) or heritability (r = 0.08). DNA markers may have only a very limited ability to predict quantitative genetic variability.

Individuals that are more heterozygous at allozyme loci are often more fit, by several measures of fitness, than individuals in the same random-mating population that are less heterozygous. Several possible explanations exist for this observation, including that allozyme heterozygosity may itself be beneficial, which implies that allozymes are not selectively neutral; allozymes may be neutral markers for chromosome segments carrying unknown genes that enhance fitness when heterozygous; or marker homozygosity may be an indicator of inbreeding depression.

Thelen and Allendorf analysed ten allozyme and ten microsatellite loci in a hatchery population of rainbow trout. Allozyme heterozygosity correlated positively with condition factor, but microsatellite, non-coding DNA heterozygosity did not. The observed relationship between heterozygosity at allozyme loci and condition factor in rainbow trout appears to be the direct effect of the allozymes themselves, rather than associative overdominance or linkage to unidentified selected genes. The results indicate that allozymes and microsatellites are differentially affected by natural selection, and that allozymes are selected whereas microsatellites are not.

Additionally, isozymes/enzymes may experience quite different rates of evolution in different species. The neutrality theory predicts that the rate of molecular evolution will be constant over time. However, the variance of the rate of evolution is generally larger than expected, according to the neutrality theory. Several modifications of the theory have been proposed to account for the 'overdispersion' of the molecular clock, such as effects due to generation time, population size, slightly deleterious mutations, repair mechanisms and more. An extensive examination of two proteins, glycerol-3-phosphate dehydrogenase (GPDH) and SOD, indicates that none of these modifications can simultaneously account for the disparate patterns observed in the two proteins. GPDH evolves very slowly in *Drosophila* species, but several times faster in mammals, other animals, plants and fungi, whereas SOD evolves very fast in *Drosophila* species and also in mammals, but much more slowly in other animals and still more slowly when plants and fungi are compared with one another or with animals. Sometimes generalizations cannot be made even with a single marker.

Genetic markers need to be matched properly with the objective of the research and the type of information and analysis being pursued. For some applications, the type of marker is not so important. For others, some markers are more powerful than others. In some cases, the type of genetic variation detected and the genetic conclusions can vary depending upon the type of marker utilized. One unfortunate result was that the lack of marker genetic variation does not eliminate the possible existence of environmentally and evolutionarily important quantitative genetic variation in the very same individuals, and this quantitative genetic variation can exist without any discernible biochemical- or molecular-marker variability.

References

Avise, J.A. (1994). *Molecular markers, natural history and evolution.* Chapman & Hall, New York

Meyer, A. (1993). Evolution of mitochondrial DNA in fishes. In: *Biochemistry and Molecular Biology of Fishes,* V. 2, Hochachka, P.W. and Mommsen, T.P. (eds). Elsevier, Amsterdam, pp 1-38

Nakamura, Y., Leppert, M., O'Connell, P., Wolff, R., Holm, T., Culver, M. and Martoin, C. (1987). Variable numbers of tandem repeats (VNTR) markers for human gene mapping. *Science,* 235: 1616-1622

Wright, J.M. (1993). *DNA fingerprinting of fishes. In: Biochemistry and molecular biology of fishes,* Hochachka and Mommsen (eds). Elsevier Science Publishers, B.V., pp. 57-91

8

Gene Mapping in Fishes

Genetics is defined as the study of heredity and embodied in the definition of heredity is the transmission of genetic characters from parents to offspring. Among the several definitions that exist for a map, the most appropriate is the concept that a map is a two-dimensional reconstruction of a part of the earth or heavens. In its totality, a genetic map does portray the ordering and landmarks of the physical entities we know as chromosomes. But in and of itself any map, including a genetic map, can only direct the observer to a specific location or place. It cannot convey to the observer the nature of that location or the activities that occur at that location. To attain an appreciation for the location we recognize that is necessary to actually visit and observe the location.

As a biological metaphor genetic maps are only the genesis for understanding the evolutionary complexities of any species. To use an analogy based upon a world map, we may regard genetic mapping as a multi-layered process spanning the identification of individual landmarks that may be miles apart on several continents, down to the cataloguing of individual grains of sand between these landmarks. By equating chromosomes to continents, landmarks to genetic markers, and grains of sand to individual nucleotide base pairs, a measure of the complexity inherent in the construction of genetic maps may be gained. Fortunately, the analogy to 'grains of sand' is somewhat of an exaggeration, as species have far fewer nucleotide base pairs than inferred. Nonetheless, the concept does portray a feeling for the magnitude of organization that is expected at each level. We have yet to witness the completion of a gene map for any vertebrate species, even though this target is nearing completion for our own species, Homo sapiens.

Studies of gene mapping in fishes are largely focused on the second level of organization (i.e., the addition of genetic markers to define linkage groups within the species). However, in the zebrafish and puffer fishes, gene mapping studies are well advanced at the highest level of organization, and a large amount of raw sequence data exists for either species in publicdatabases.In addition, plans are currently underway to sequence the entire genome of the puffer?sh due to its relatively small size (~400 Mb) which would make it the first teleost for which complete genome information is available. It is expected that functional gene coding regions will be in a higher density in the pufferfish per number of contigs assembled and thus sequencing efforts are expected to yield more immediate rewards.

Choice of Markers

Segregation analysis of polymorphic markers allows the assignment of isozymes, biochemical markers, DNA fragments and genes to chromosomes, ordering of these genetic markers and genes, establishment of genetic linkages, mapping of important genes and identification of modes of inheritance of production traits by demonstrating the linkage of major genes, either singly or as part of polygenic systems. Initially, most of this research was conducted with isozymes and easily scored qualitative traits; however, advances in DNA technology during the past decade have allowed rapid identification of very large numbers of markers for producing genetic linkage maps—the mapping of genes on chromosomes. Isozymes and other biochemical markers, such as genes associated with the immune system, have the probable advantages of being conserved across taxa and linked to quantitative traits, but have the disadvantage of being few in number. However, the advent of ESTs is rapidly overcoming this shortcoming for type I markers—actual genes—for gene mapping.

Assuming a recombination genome size to be 2000 centimorgans (cM), 200 evenly distributed genetic markers would be required for a map with a resolution of 10 cM (or a map for any locus being within a 5 cM average to the nearest marker). Theoretically, a genetic map with 1000 molecular markers can place any locus in close proximity to less than a million base pairs. Thus, several thousands of markers may be needed to construct genetic linkage maps that can tightly localize specific genes for practical applications, such as gene isolation.

Microsatellites and AFLP markers are the most reliable, efficient and abundant markers for detailed genetic linkage mapping in catfish, and perhaps other aquatic organisms. Fine linkage mapping depends on the availability of large numbers of ESTs and the anchoring of well-ordered contigs of bacterial artifical chromosome (BAC) clones to linkage maps.

Conservation of microsatellites across a broad range of taxa in aquatic organisms has potentially important applications and implications for gene mapping, marker-assisted selection (MAS), cloning of genes and evolutionary studies. High levels of genetic conservation allow comparative gene mapping, which is especially important considering the rare availability of type I markers (markers that encode genes) in fish (Z.J. Liu, unpublished results). Comparative gene mapping would facilitate rapid advancements in gene mapping of major aquaculture species.

Conservation of microsatellite loci across a broad range of species has been demonstrated among various taxa. In the majority of these studies, primers designed from microsatellite-flanking regions of one species were successfully evaluated in closely related species. For example, primers from humans were tested in other primates, or primers from one species member of a family were evaluated for their conservation among other species in the family.

Microsatellites are highly conserved among species of fish. Most channel catfish, Ictaluridae, primers amplified microsatellites from Cichlidae. Homologous microsatellite loci have endured for approximately 300 million years in turtle and for about 470 million years in fish. Microsatellite-flanking sequences of fish may evolve at a slower rate than those of mammals based on fish gene maps generated to date. The identification of homologous chromosome segments in siluriform, cyprinodontiform and salmoniform fishes supports the hypothesis that teleostean gene arrangements may have diverged more slowly from those of the vertebrate ancestor than those same gene arrangements in mammalian orders. Gene map locations in one teleost may be highly predictive of map locations in other fish.

Mapping Systems

Gene mapping can be accomplished by using either intraspecific or interspecific systems. Interspecific hybridization systems are powerful for the construction of genetic linkage maps. Interspecific approaches to gene mapping have frequently been used, in recent years utilizing both microsatellites and AFLPs and in the early fish gene maps generated with isozymes because of the power and high level of polymorphism found in interspecific systems. Liu *et al.* utilized a channel catfish **X** blue catfish hybrid system for gene mapping. The F1 hybrids are fertile and F_2 hybrids or back-cross progeny can be readily produced. The primary advantages of using this hybrid system for genetic linkage analysis are the high level of polymorphism between channel catfish and blue catfish and the fact that few reference families are needed because single families can be generated that are heterozygous for virtually every genetic marker. Additionally, there are large performance differences between species for production traits, which could expedite QTL mapping.

Conservation of microsatellite loci between closely related species, as demonstrated by Liu *et al.* for ictalurid catfish, allows construction of unified maps among those generated from intraspecific and interspecific mapping systems. One individual channel catfish was heterozygous for 24 of 31 microsatellite loci. Similar results were generated for blue catfish, white catfish and flat-head catfish. These results indicate that small numbers of interspecific or intraspecific reference families would be sufficient for gene mapping of microsatellite loci, and microsatellite markers would be almost as numerous as RAPD and AFLP markers. The microsatellites would, in fact, be even more powerful because of their codominance and ability to identify heterozygotes.

Utilization of haploid gynogenesis is another powerful mapping strategy. Single sperm typing is now possible but haploid gynogenesis has advantages compared with single sperm typing. Cell division in eggs is activated by irradiated sperm, resulting in haploid individuals, representing a single maternal meiotic event with no paternal genome contribution. The haploid gynogenesis strategy represents the female counterpart of sperm typing but has the advantages that no individual sorting is needed, repeated tests are possible on the same individual because of the large amount of cells and DNA, it is not entirely dependent on PCR and markers that can be mapped are not restricted to non-coding nuclear DNA and include isozymes, mtDNA and ESTs. Both single sperm typing and haploid gynogenesis allow resolution of recombination rates below 0.5% and discrimination of loose linkages above 45%, up to no linkage, 50%. Computer simulation indicates that lethal genes, which may eliminate specific haplotypes and cause segregation distortion of markers linked with such genes, do not interfere with the recombination estimate. Additionally, haploid gynogenesis indicates the location of putative lethal genes relative to the informative markers. This strategy is efficient in distinguishing between variants, and allowed Lie *et al.* to detect segregation distortion of microsatellites in Atlantic salmon, probably due to pre-selection of eggs or embryos resulting in differential mortality of certain genotypes.

LINKAGE DISEQUILIBRIUM

When conducting linkage or population genetic analyses, linkage disequilibrium is sometimes observed. Linkage (or gametic) disequilibrium is when there is a lack of fit for observed two-locus gametic frequencies compared with those expected, based on the product of the single-locus allelic frequencies. The frequency of an $A1B_2$ gamete (loci A and B) in the population should be equal to frequency of the Al allele times the frequency of the B_2 allele.

Linkage disequilibrium should decay by $1-r$ each generation for random mating, where r is the recombination rate between the two loci. The value for r can vary from zero for complete linkage to 0.5 for no linkage; therefore, linkage disequilibrium will decay by one-half each generation for most pairs of loci.

Linkage disequilibrium can be caused by mixtures in the sample of two or more populations with different allelic frequencies, a founding population already in disequilibrium, selection for certain heterozygous genotypes or random genetic drift to high frequencies of particular chromosome types. In most population studies, the linkage disequilibrium is caused by mixing or founder effects. If a population has a bottleneck with a low effective population size, linkage disequilibrium might be expected for several generations.

For most population studies, variance components of the linkage disequilibrium values will only help indicate recent mixing (zero to two generations) of two highly divergent intraspecific gene pools. Forbes and Allendorf found linkage disequilibrium values for linkages that have not yet decayed through recombination for fixed alternate alleles for several mixed populations of two distinct subspecies of cutthroat trout, Westslope and Yellowstone, after five to 15 generations of interbreeding.

Isozyme Maps

Isozymes were the first biochemical or molecular markers utilized for gene mapping in fish. Pasdar *et al.* examined linkage relationships of nine enzyme loci—aconitase (ACON), esterase (EST), glucose-phosphate isomerase A and B (GPI), glycerate-2-dehy-drogenase (G2DH), malic enzyme (ME), phosphoglycerate kinase (PGK), phospho-glucomutase (PGM) and superoxide dismutase (SOD)—in backcrosses of reciprocal F_1 hybrids between green sunfish *(Lepomis cyanellus)* and red-ear sunfish *(Lepomis microlophus)* to each of their two parental species. A three-point linkage map containing G2DH, PGK and SOD was obtained, with frequencies of recombination between G2DH and PGK and between PGK and SOD at 45.3 and 24.7%, respectively. The remaining six loci assorted independently.

Although partial tetraploidy is widespread in the teleost genome, salmonids present one of the most obvious and most studied problems in fish gene mapping because of their tetraploid ancestry. Johnson and Wright and Johnson *et al.* examined the joint segregation analyses of isozyme loci in males and females of seven species and three fertile species hybrids of trout, charr and salmon, and identified 15 linkage groups. Johnson and Wright were proponents of the interspecific approach to gene mapping, and concluded that, since linkage groups are highly conserved among species and hybrids, they can be combined to form a common linkage map, in this case one for salmonids. Pseudolinkage was observed and was explained as preferential multivalent pairing and disjunction of metacentric (centrally fused) chromosome arms with homoeologous arms of other chromosomes in male salmonids, in contrast to bivalent pairing in females. Five pseudolinkage groups were detected and were highly conserved among species and hybrids. Conservation of pseudolinkages among salmonids must have been a result of major chromosomal fusions in a common tetraploid ancestor before the radiation of salmonid species.

Earlier, Davisson *et al.* and Lee and Wright had observed the psuedolink-age found in some salmonid males. Both cytological and linkage analyses indicated that spontaneous centric fusion and fission could account for the curious patterns of pseudolinkage of two lactate dehydrogenase (LDH) loci in males of brook trout and in the F1, F_2 and backcross generations of lake trout X brook trout hybrids. Intraindividual polymorphisms for acrocentric and metacentric chromosomes in somatic and gonadal tissue of these fish are consistent with the proposed polyploid evolution in Salmonidae. Mitotic and meiotic analyses of the tetraploid-derivative species, brook trout, indicated that the process of diploidization was incomplete.

The diploid number was 2N = 84, with 16 metacentrics and 68 acrocentrics for both males and females from different sources, and no inter- or intraindividual Robertsonian polymorphism was present. Oocytes at pachytene had the expected 42 bivalents with eight metacentric and 34 acrocentric pairs. However, variable numbers of tetravalents, with a total of 35-40 bivalent plus tetravalent elements, were observed in metaphase I cells of males. Each tetravalent was composed of two acrocentric and two metacentric chromosomes. Variability in the number of tetravalents was found not only among different brook trout sources, but also among different cells of the same fish. Differential homoeologous pairing was proposed to account for the variable number of tetravalents and to explain the occurrence of pseudolinkage in some salmonid males.

Some of the linkage relationships in salmonids are due to duplicated loci resulting from the tetraploid ancestry. Hollister *et al.* observed variable genotypes for the duplicate loci encoding the enzyme peptidase D (PEPD) in lake trout, brook trout and their fertile hybrid (splake). Non-random assortment was observed among progeny of parents doubly heterozygous for the PEPD-1 and PEPD-2 loci, the duplicate loci encoding GPI and the locus sorbitol dehydrogenase (SDH). Linkage groups were PEPD-1 with GPI-1 and PEPD-2 with GPI-2 with SDH. The results fitted and were consistent with the earlier-determined chromosomal model involving preferential tetravalent pairing of homoeologous chromosomes—pseudolinkage.

Disney and Wright later observed extensive multivalent pairing in lake trout, which, along with data on hybrid splake (brook X lake) trout, supported a meiotic model to explain pseudolinkage. Additionally, C-banding of mitotic and meiotic lake trout chromosomes revealed an intraindividual polymorphism for a Robertsonian fusion, and silver staining showed that the chromosomes with active nucleolar organizer regions located proximal to a centromere were not involved in the fusion event.

Null alleles can also complicate linkage analysis in salmonids and in oysters. Unusual phenotypic distributions were observed at the muscle-specific, duplicate aspartate aminotransferase (AAT) locus in a wild population of brook trout, and analysis of these phenotypic distributions eliminated disparate gene frequencies, non-random association between the two loci and inbreeding as possible explanations. Models incorporating a null allele and inheritance data from hatchery populations of brook trout fitted the data and confirmed a null-allele polymorphism. This AAT null allele, along with other null-allele polymorphisms in salmonids, is evidence that loss of duplicate gene expression is still occurring; however, there is no such evidence of ongoing loss of duplicate gene expression in the Catostomidae, another tetraploid-derivative lineage.

Early gene-mapping research was also conducted with *Xiphophorus,* utilizing interspecific approaches. Morizot *et al.* obtained a three-point linkage group comprised of loci coding for adenosine deaminase (ADA), glucose-6-phosphate dehydrogenase (G6PDH) and 6-phosphogluconate dehydrogenase (6PGD) for *Xiphophorus* (Poeciliidae) by utilizing reciprocal backcross hybrids from crosses between either *Xiphophorus helleri guentheri* or *X. helleri strigatus* and *Xiphophorus maculatus.* The alleles at this linkage group assorted independently from the alleles at isocitrate dehydrogenase (IDH) 1 and 2 and glyceraldehyde-3-phos-phate dehydrogenase (GAPDH) 1, and the latter assorted independently from each other. The linkage group was conserved in all populations of both species of *Xiphophorus* examined. Data from *X. helleri guentheri* backcrosses indicate the linkage relationship, ADA-6%-G6PDH-24%-6PGD, and ADA-29%-6PGD (30% when corrected for double crossovers), but results from backcrosses *X. helleri strigatus* gave different recombination frequencies for the same gene order. Possible explanations include differences due to an inversion or a sex effect on recombination. The linkage of 6PGD and G6PDH exists in species of at least three classes of vertebrates.

Recombination data from backcross hybrids among three species and four subspecies of *Xiphophorus* indicate four additional linked loci, linkage group (LG) II, (esterase) EST-2-0.43-EST-3-0.26- (retinal lactate dehydrogenase) LDH-1-0.19-

(man-nose phosphate isomerase) MPI. Interference was detected in the EST-3 to MPI region, and LG II assorted independently from the six loci of LG I and from GAPDH-2 and IDH-2.

Next, Morizot *et al.* analysed 76 polymorphic isozyme loci in backcross hybrid individuals from intra- and interspecific crosses of the genus *Xiphophorus* (Poeciliidae), identifying 17 multipoint linkage groups containing 55 protein-coding loci and one sex-chromosome-linked pigment-pattern gene. Gene orders were determined for ten linkage groups, and total genome length was estimated to be 1800 cm.

Comparisons of the *Xiphophorus* linkage map with those of other fishes, amphibians and mammals suggested that fish gene maps are remarkably similar and probably retain many syntenic groups.

Six isozyme linkage groups have been established for ictalurid catfish, using the channel-blue catfish interspecific hybrid system. Gene-centromere distances were estimated for six loci in gynogenetic channel catfish and for additional polymorphic loci in blue-channel triploid hybrids. At least 28 polymorphic isozyme loci were found and used to establish channel catfish multipoint LGs I-VI, comprising 18 loci (R.A. Dunham, B. Argue and D. Morizot, unpublished). Eleven unlinked loci may bring the total of isozyme-marked chromosomes to 17 of the 29 chromosome pairs of channel and blue catfish. The extensive genetic variability within and between blue and channel catfish at approximately 70 isozyme loci could allow significant expansion of this isozyme-based gene map. *Ictalurus* LG I is comprised of loci coding for glutathione reductase and PGM. Three other loci assort independently from the LG I pair, providing isozyme markers for four chromosomes.

Three isozyme loci assigned to catfish linkage group LG II are also linked in LG II of poeciliid fishes, indicating the evolutionary conservation of both neutral and physiological markers. Comparison of the gene maps of poeciliid and salmonid fishes substantiates that this result is not a rare event. Orthology is sometimes difficult to establish because poeciliids are diploid and salmonids are recently tetraploid-derived.

However, at least four cases of linkage-group conservation have been identified, and linkage-group divergence has yet to be observed, again illustrating the strength of the linkage conservation in fish. Within poeciliids, LGs II of *Xiphophorus* and *Poeciliopsis* are homologous and are homologues of *Poecilia* LG I. Additionally, poeciliid and salmonid syntenies can be identified in centrarchid fishes and ranid frogs and are apparently even evolutionarily conserved in segments of human chromosomes 12, 15 and 19. Morizot and Morizot *et al.* proposed that the chromosomal arrangements of duplicated genes in fishes suggest retention of patterns produced by three rounds of tetraploidization.

In *Xiphophorus* poeciliids, CA1 (orthologous to catfish CAH-2) has been assigned to LG XXIV (the *X. maculatus* sex-chromosome linkage group), and GAPDH-2 (orthologous to catfish GAPDH-1) is linked to PEPC (orthology with catfish dimeric peptidases is uncertain) in LG U3. Sex-chromosome linkage groups vary widely even within orders for fish.

Catfish LGs II and III are strong examples of evolutionary conservation of linkage groups in fishes and between fishes and mammals. The orthologues of the three catfish LG II isozyme loci are also linked in *Xiphophorus* LG II, and GPI and MPI orthologues are also linked in salmonid LG 13. Orthologues of MPI and a-mannosidase are syntenic on human chromosome 15. Catfish LG III shows homology with other fish linkage groups. *Xiphophorus* LG IV contains pyruvate kinse (PK), GPI, PEPD and probably a cytosolic IDH locus, which is orthologous to the catfish arrangement of loci. Salmonid LG 3 also contains GPI and PEPD loci, as well as human chromosome 19.

The sole example of definitive divergence between catfish, poeciliid and salmonid linkage groups involves muscle IDH (mIDH). The location of mIDH in catfish LG III is different from that in poeciliids and salmonids, and the orthologies of PNP and EST loci in catfish LG III with those of loci in other fishes are not yet absolutely determined. mIDH is linked to LDH-A, LDH-C and two esterase loci in *Poeciliopsis* LG I, which is homologous to *Xiphophorus* LG II. This linkage is conserved in salmonid LG 13, which is also partly homologous to

Xiphophorus LG II. Loci for muscle PK, mIDH and MPI appear to be conserved on human chromosome 15, suggesting a symplesiomorphic vertebrate gene arrangement. In contrast, the catfish position of mIDH in LG III is within a segment conserved in *Xiphophorus*

LG IV, salmonid LG 3 and human chromosome 19. The apparent homology of catfish LG III with synteny of GPI, PEPD and mIDH on mouse chromosome 7 might suggest a polymorphic symplesiomorphy Morizot proposes that linkage of duplicates of GPI, IDH, PK and, by extension from salmonids, AAT and malate dehydrogenase (MDH) loci indicates that poeciliid LGs II and IV (catfish LGs II and III) are derived from ancestral chromosome duplications and translocations, which are more likely to occur between homoeologous chromosomes, perhaps because of mispairings in regions of residual homology. The translocations of mIDH on to catfish LG III and mouse chromosome 7, therefore, both represent a relatively frequent type of chromosomal rearrangement.

Since chromosome numbers vary widely among fish, some linkage-group divergence among fish gene maps should be expected, but the observed example differs from the expected pattern of linkage-group splitting to produce new linkage groups. The loci assessed for linkage in ictalurid catfish include markers assigned to multipoint linkage groups in poeciliid and/or salmonid fishes. These studies indicate that many, but not all, gene arrangements among fish gene maps can probably be predicted from the gene arrangements of other fish gene maps. While no mechanisms have been found that result in gene-arrangement conservation, it is possible that highly conserved areas of the genome may be under stringent selection, and that disruption of these arrangements may result in disturbances of gene regulation.

Gene sequencing has now begun to confirm these hypotheses and observations from gene-mapping studies. Ninety-five per cent of the *Fugu rubripes* genome has been sequenced and, as in the human genome, gene loci are not evenly distributed but are clustered into sparse and dense regions. Some 'giant' genes had average coding-sequence sizes, but were spread over genomic lengths significantly larger than those of their human orthologues. Three-quarters of predicted human proteins match strongly to *Fugu,* but approximately one-quarter have highly diverged from *Fugu* or have no pufferfish homologues, illustrating the extent of protein/gene evolution in the 450 million years since teleosts and mammals diverged. Some *Fugu* genes, either specific to teleosts or lost in human lineages, are no longer found in humans. In general, conserved linkages of chromosomal segments were preserved from the common vertebrate ancestor of *Fugu* and humans, but with considerable scrambling of gene order. Additionally, several conserved non-coding sequences in the promoter and intronic regions were identified that are probably involved in gene regulation. More than 80% of the gene assembly of *Fugu* was multigene-sized scaffolds. Repetitive DNA accounted for less than one-sixth of the sequence, and gene loci occupied about one-third of the genome in this small 365-megabase genome.

The evolution of mitochondrial genomes is not necessarily parallel to that of the nuclear genome. The gene order, nucleotide composition and evolutionary rate of the mtDNA genome of *Fugu* correspond to those of other teleosts, suggesting that the evolution of this genome was not affected by the processes that led to the dramatic reduction of the size of the nuclear genome of *Fugu.*

Divergences of catfish and *Xiphophorus* occur for gene arrangements where human and *Xiphophorus* gene arrangements are also dissimilar and where rearrangements are found in human or rodent lineages. These chromosomal segments provide a focal point for critical testing of homology of fish gene arrangements in three teleost orders, involving gene families with members located on human chromosomes 11, 12, 15 and 19.

Human chromosome 19 contains the most highly conserved genetic linkage in vertebrates, the GPI and PEPD loci. Synteny of these genes has been shown for humans, a variety of other mammals, amphibians, salmonid fishes and the poeciliid fishes *Xiphophorus* and *Poeciliopsis.* Not a single case of asynteny has been found among vertebrate gene maps.

A second strongly conserved DNA segment is homologous to arrangements on human chromosome 15 and includes PKM2 and MPI, which are conserved in poeciliids, salmonids and mammals. The only exception to the linkage of these two genes so far identified in mammals is in bovines, where PKM2 is located on chromosome 10 and MPI on chromosome 21. Linkage groups of fishes homologous to these segments of human chromosomes 15 and 19 appear to have originated by chromosome duplication, as evidenced by the linkage of GPI, IDH and PK loci in each linkage group. The presence of duplicate genes for cytosolic and membrane-bound isozymes of a-man-nosidase on human chromosomes 15 and 19 is another example of this pattern of linkage of gene duplicates.

Surprisingly, the first definitive case of linkage-group divergence among teleost gene maps involved two linkage groups that are otherwise highly conserved throughout vertebrate evolution. Gene maps of mammals other than mouse and human contain numerous examples consistent with the hypothesis that most translocations occur between paralogous chromosomal segments. For example, the most common location of NP in mammalian gene maps appears to be in syntenic groups homologous to human chromosome 15 segments. Synteny of purine nucleoside phosphorylase (NP) with creatine kinase (CK) B, PKM2 and usually MPI occurs in many mammalian species. However, NP and CKB are syntenic on chromosome 14 in humans, while in apes NP and CKB are asyntenic. Rabbits have diverged even further not only by having NP asyntenic with CKB, but also by relocating NP on to yet another paralogue, syntenic with PEPB, which is found on human chromosome 12, syntenic with LDH-B, an arrangement also exhibited by salmonid fishes. Mechanisms must exist that make certain chromosome segments across orders susceptible to rearrangements and translocations.

Linkage of LDH-A (human chromosome 11) to human chromosome 15 loci, mIDH, PK2 and MPI in poeciliid fishes, similar to synteny of mIDH and LDH-A in frogs, further illustrates this cycle of translocations among paralogous segments of human chromosomes 11, 12, 15 and 19, which are predicted to have arisen through a minimum of two rounds of tetraploidization. The presence of mIDH in poeciliids on a human chromosome 15 homologue and in catfish in a human chromosome 19-related syntenic group is logical, as exactly the same pattern is evident in mammals by the location of sorbitol dehydrogenase (SORD) on human chromosome 15 syntenic with GPI and PEPD (human chromosome 19) in African green monkeys.

For practical genetic improvement for aquacultural species, understanding why, how and in what directions linkage arrangements have diverged is less important than the predictiveness implied by slow divergence of teleost gene arrangements from those of the vertebrate ancestor. This should allow research with one species usually to be of practical application in another species. To provide the most conclusive estimates of rates of gene-map divergence in fishes, the segments where fish and mammalian gene maps have substantially diverged should be examined. The comparison of catfish gene arrangements of orthologues located on human chromosome 19 with those mapped in *Xiphophorus* illustrates the clearest test of teleost linkage-group conservation. Chromosome 19 in human carries loci of GPI, PEPD, CKM (= CK-A), ERCC2, LIGI and CMT genes (another possible orthologue, PGK2, is listed in many human gene-map summaries, but in fact is a pseudogene (D.C. Morizot, personal communication)). In *Xiphophorus,* GPI and PEPD loci are linked in LG IV (and in catfish LG III), but, in contrast to humans, ERCC2 and CK-A are assigned to LG XIV and LIGI to LG VI and CMT is polymorphic but is as yet unassigned to the gene map.

Loci for muscle PK, mIDH and MPI appear to be conserved even further on human chromosome 15, suggesting a symplesiomorphic vertebrate gene arrangement. The catfish position of mIDH, in contrast, is within a segment conserved in *Xiphophorus* LG IV, salmonid LG 3 and human chromosome 19. The apparent homology of catfish LG III with synteny of GPI, PEPD and mIDH on mouse chromosome 7 might suggest a polymorphic symplesiomorphy; however, the following explanation proposed by Morizot seems more likely. Linkage of duplicates of GPI, IDH, PK and, by extension from salmonids,

AAT and MDH loci indicates that poeciliid LGs II and IV (catfish LGs II and III) are derived from ancestral chromosome duplications, and translocations are more likely to occur between homoeo-logous chromosomes, perhaps because of mispairings in regions of residual homology. The translocations of mIDH on to catfish LG III and mouse chromosome 7, then, represent examples of a relatively frequent type of chromosomal rearrangement.

DNA Markers and Maps

The development of powerful DNA techniques, such as AFLP, RAPD and microsatellites, which generate large numbers of molecular markers, has allowed rapid progress in aquatic organism linkage mapping. Construction of framework genetic linkage maps has progressed rapidly in aquaculture species such as catfish, salmonids, shrimp and oysters, and espe cially for model species, such as zebra fish and *Fugu*. Radiation hybrid panels have been developed in tilapia, BAC libraries in catfish (G.C. Waldbieser, unpublished data) and normalized complementary DNA (cDNA) libraries constructed for EST analysis and functional genomics analysis.

Over 350 microsatellite markers and over 600 AFLP markers have been mapped for channel catfish. Outbred populations of channel catfish contained an average of eight microsatellite alleles per locus and an average heterozygosity of 0.70. A total of 293 microsatellite loci were polymorphic in one or two families, with an average of 171 informative meioses per locus and these data were used to construct a genetic linkage map of the channel catfish genome (N = 29). Nineteen type I loci, 243 type II loci and one EST were placed in 32 multipoint linkage groups covering 1958 cM. Nine additional type II loci were contained in three two-point linkage groups covering 24.5 cM, 22 type II loci were unlinked and multipoint linkage groups ranged from 11.9 to 110.5 cM, with an average intermarker distance of 8.7 cM.

In the case of the interspecific mapping strategy of Liu *et al.*, a total of 607 polymorphic AFLP loci were produced, with the 64 *EcoRI/MseI* primer combinations, using a (channel catfish female X blue catfish male) X blue catfish male backcross family. A total of 101 markers (16.6%) were not used for the construction of the linkage map because they showed distortion—linkage disequilibrium—from the expected 1:1 ratio. Four hundred and forty-five markers were assigned to 40 linkage groups, with 29 markers unlinked, and 32 markers were excluded because of large map distances. The number of markers on the 40 linkage groups ranged from two to 33, compared with two to seven markers per linkage group for fish isozyme maps, 13 to 49 for various types of markers, but primarily AFLPs, in medaka, *Oryzias latipes*, and 73 to 201 ESTs for extensive zebra fish, *Brachydanio rerio*, maps. There were 25 major linkage groups with five to 33 AFLP markers and 15 small linkage groups with two to four AFLP markers with genomic coverage of this AFLP linkage map spanning 2511 cM Kosambi, compared with the 1958 cM channel catfish map based on microsatellites.

The largest linkage group spanned 216.9 cM, with 19 AFLP markers, and the smallest 0.0 cM, with two AFLP markers, with the mean for the linkage map of one AFLP marker every 5.6 cM. However, the distances between any two given markers varied greatly, ranging from 0 cM to 51.2 cM. The microsatellite catfish map was less variable in regard to range of length of linkage groups, with multipoint linkage groups ranging from 11.9 to 110.5 cM, with an average intermarker distance of 8.7 cM, similar to the mean intermarker distances for the AFLP map. Eight of the AFLP linkage groups were longer than the longest microsatellite linkage group, indicating that in many cases the microsatellite strategy may be underestimating the size of the linkage group, which may be related to marker distributions. Karyotypes of channel and blue catfish are indistinguishable and both species have small chromosomes. However, some channel catfish chromosomes are larger than those of blue catfish, with length ranging from 1.5 to 3.5 μm and 1.0 to 2.0 μm for channel and blue catfish chromosomes, respectively. In both species, the relative size of the individual chromosomes ranges from 2 to 10% of the total complement length. Based on chromosome length for channel catfish, the largest linkage group would be expected to be approximately

2.3 times larger than the smallest linkage group. Future fine-tuning of the catfish map and consolidation of the linkage groups should bring the variation in linkage length closer to this much tighter range, compared with the much more variable length of the microsatellite and AFLP catfish linkage maps.

Based on segregation ratios in 41 haploid embryos derived from a single *Oreochromis niloticus* female, a tilapia gene map was developed from 62 microsatellite and 112 AFLPs. Linkages were identified for 162 of the markers, with 95% of the microsatellites and 92% of the AFLPs linked on the map. The map covered 704 cM Kosambi in 30 linkage groups covering the 22 chromosomes of Nile tilapia and 24 of these linkage groups contain at least one microsatellite polymorphism. From the percentage of markers 15 or fewer cM apart, a total map length of 1000-1200 cM was estimated. High levels of interference were observed, consistent with measurements in other species of fish. The *Oreochromis* gene map now contains more than 500 microsatellites, indicating that the genome size is about 1 gigabase and the map length is 500 kb/cM.

Agresti *et al.* utilized the interspecific approach to develop tilapia gene maps. A synthetic stock (artificial centre of origin (ACO)) was produced by crossing five groups of fish, *O. niloticus* (wild type (On) and red (ROn) strains), *Oreochromis aureus* (Oa), *Oreochromis mossambicus* (Om) and *Sarotherodon galilaeus* (Sg). Three-way cross families (3WC) and four-way cross families (4WC) were produced, to introgress all four species into the ACO. An Om **X** (Oa **X** ROn) family was used to develop a gene map using microsatellite and AFLP DNA markers. The female (Om) parent had a total of 78 segregating markers (17 microsatellites, 61 AFLPs), and 62 markers (13 microsatellites, 49 AFLPs) were linked in 14 linkage groups covering a total of 514 cM. The F_1 hybrid male parent had a total of 229 segregating markers (62 microsatellites, 167 AFLPs), and 214 of these markers (60 microsatellites, 154 AFLPs) were linked in 24 linkage groups covering a total of 1632 cM.

Young *et al.* conducted gene mapping for rainbow trout by utilizing doubled haploids—androgens. The sex-determining locus was at a distal position on one of the chromosomes.

AFLPs appeared to be primarily clustered near centromeres, variable-number tandem repeats were frequently more telomeric and salmonid-specific small interspersed nuclear elements were intermediate in distribution compared with the other two marker types. This is another example of the choice of markers affecting the outcome of a genetic analysis.

Paint probes of whole-arm chromosomes of rainbow trout were specific for single pairs of arms, suggesting that the majority of chromosomes from ancestral tetraploids have diploidized. Paint probes for the sex chromosomes of lake trout paint different autosomal chromosome pairs for rainbow trout and chinook salmon, indicating separate evolution of sex chromosomes in *Salvelinus* and *Oncorhynchus*. Similar fluorescent techniques have been used for gene mapping in salmonids. The Y chromosome in chinook salmon has been identified using fluorescence *in situ* hybridization (FISH), with a probe to a male-specific repetitive sequence isolated from this species. The probe lights up the distal end of the short arm of an acrocentric chromosome.

Sakamoto *et al.* constructed a rainbow trout—a tetraploid-derivative species -gene map with 191 microsatellites, three RAPD, seven expressed sequence marker polymorphisms (ESMP) and seven allozyme markers. Twenty-nine linkage groups were identified, with potential arm displacement in the female map because of male-specific pseudolinkage arrangements. Synteny of duplicated microsatellite markers confirms some previously reported pseudolinkage arrangements based upon allozyme markers. Fifteen centromeric regions were identified with a half-tetrad analysis using gynogenetic diploids. Female map length was about 1000 cM, but this is a substantial underestimate, as many genotyped segments remained unassigned at the LOD threshold of 3.0. Female:male map distances were tremendously different. Females had an eightfold lower recombination rate in telometric regions compared with males, whereas proximal to the centromeres the female recombination rate was tenfold greater than that of males. Sakamoto proposed that quadrivalent formations, which were almost exclusive in males, is the cause of the sex-specific differences in recombination rates.

The initial linkage map of zebra fish, *Brachydanio rerio,* was constructed using haploid genetics and was saturated with 401 RAPD markers and 13 simple sequence repeats, spaced at an average interval of 5.8 cM. This strategy allowed rapid mapping of lethal and visible mutations. Then Johnson *et al.* expanded the zebra-fish map using half-tetrads. A total of identified 652 PCR-based markers were closely linked to each of the 25 centromeres of zebra fish.

Since then, radiation hybrid maps have been generated for zebra fish containing 4226 markers, including 459 genes and 3867 ESTs, covering more than 88% of the genome. The map was 14,372 centirays, and average breakpoint frequency corresponded to 1 centiray = 118 kb. The concordance between radiation hybrid maps and meiotic maps was 96%. The distribution of the ESTs was very uniform, ranging from 73 to 201 ESTs per linkage group, indicating that there are no gene-rich or gene-poor chromosomes in zebra fish.

The zebra-fish map now contains more than 1267 expressed sequences. Postlethwait *et al.'s* data indicate that in some cases the entire content of some human chromosomes has been conserved for the 450 million years since the lineages of zebra fish and humans diverged. Intrachromosomal rearrangements have been frequent, resulting in altered gene orders within conserved syntenies. The maps indicate that about 30% of the zebra-fish genes have been retained from a genome duplication that probably occurred before teleost radiation. Teleost and mammalian genomes have about the same number of chromosomes despite the teleost genome duplication, probably because of fusions of human chromosomes not because of fusions of fish chromosomes. Zebra-fish chromosomes can be identified by AT-rich repetitive sequences at the centromere and GC-rich sequences adjacent to the centromeres.

Wada *et al.* found 227 informative RAPD markers in the Japanese medaka, *Oryzias latipes,* segregating at 170 loci, including three pigment-pattern loci, five enzyme-coding loci and one male-determining factor. A gene map was constructed consisting of 28 linkage groups spanning about 2480 contiguous cM with an average of 323 kb/cM. Naruse *et al.* used a reference-typing DNA panel from 39 cell lines derived from backcross progeny to map 633 markers (488 AFLPs, 28 RAPDs, 34 IRSs, 75 ESTs, four sequence-tagged sites (STSs) and four phenotypic markers) for the medaka, *O. latipes,* of the order Beloniformes. The total map length of 24 linkage groups—the haploid number for medaka—was 1354.5 cM, and 13–49 markers were obtained for each linkage group. Conserved synteny between medaka and zebra fish was observed for two independent linkage groups; however, unlike zebra fish, the medaka linkage map exhibited obvious restriction of recombination on the linkage group containing the male-determining region (Y) locus compared with the autosomal chromosomes. Different genomic phenomena can be seen among different species of fish.

Gene mapping in Pacific oysters and eastern oysters is difficult because of distortion of Mendelian segregation ratios, and this might be aggravated by a high frequency of null alleles. Hedgecock found by utilizing microsatellite markers in F_2 crossbreeds that this distortion is caused by a very large number of deleterious recessive alleles/mutations carried by oysters.

Wilson *et al.* used 673 AFLP polymorphic markers and some microsatellites that were free from segregation distortion to develop a low-density linkage map for black tiger shrimp, *Penaeus monodon. A* total of 116 markers segregated in more than one of the experimental families, resulting in 20 distinct linkage groups covering a total genetic distance of 1412 cM, although black tiger shrimp have more than 40 chromosome pairs.

Major Histocopatibility Complex and Oncogenes

Genes other than isozymes and DNA markers have recently been mapped in fish, such as major histocompatibility complex (MHC) class genes and tyrosinase genes, which, in the future, may serve as good starting-points to search for related QTLs. McConnell *et al.* isolated two MHC class II B genes from an inbred *X. maculatus* strain, Jp 163 A, and mapped one of these genes, *DXB,* to linkage group III, linked to a malic enzyme locus, which is also syntenic with human and mouse MHC. A second type of class II B clone, a DAB-like gene, was orthologous to class II genes identified in other fishes, and was 63% identical to the *X. maculatus* DXB sequence in the conserved B_2-encoding exon. This second *DXB* gene is an unlinked duplicated locus not previously identified in teleosts, and was assigned to a new linkage group, LG U24.

Interspecific hybrids of *Xiphophorus* have a number of simple oncogenes and tumour-suppressor genes, and these fish serve as malignant melanoma models. The gene map of *Xiphophorus* has about 100 genes assigned to at least 20 independently assorting linkage groups, in addition to more than 250 anonymous DNA sequence markers. Morizot *et al.* mapped the tumour-suppressor locus, DIFF, which is one of two genetic determinants of melanoma formation in the hybrid melanoma, the Gordon-Kosswig melanoma model. The other gene responsible for melanoma formation in this model is a sex-linked tyrosine kinase gene, *Xiphophorus* melanoma receptor kinase *(Xmrk),* which is related to *EGFR.* The cellular oncogene homologues of the non-receptor tyrosine kinase family orthologous to yes and fyn are over-expressed in malignant melanomas of *Xiphophorus* and may be involved in tumour progression. The *Xiphophorus* yes gene, *YES1,* belongs to LG VI, closest to the *EGFR* gene, and a *fyn* gene homologue to LG XV, linked to the gene for cytosolic a-galactosidase. The *EGF-R*-related sequence (EGFRL1) previously assigned to *Xiphophorus* LG VI was determined to be the *EGFR* orthologue. Morizot *et al.* conclude that the presence of expressed duplicates of members of the tyrosine kinase gene family in teleost fishes may increase the potential number of targets in oncogenic cascades in fish tumour models.

Wellbrock *et al.* reviewed the genetics of melanoma in *Xiphophorus* and indicate that the primary event for tumour formation is the cell lineage-specific overexpression of a structurally altered tyrosine kinase receptor. This phenomenon is also found in many tumours of birds and mammals. Once expressed at high levels, the *Xiphophorus* melanoma inducing receptor kinase *Xmrk* shows constitutive activation. Analyses of the different signalling cascades induced by the Xmrk receptor has led to the identification of the src-kinase Fyn, the MAP kinases ERK1 and ERK2, the signal transducer and activator of transcription (STAT5) and the PI3-kinase as its major downstream substrates.

EFFECTS OF KARYOTYPES, CLUSTERING AND DISTORTION

Theoretically mapping results can be affected by the karyotype, DNA profile and chromosome characteristics of the species of interest. The mapping results of Liu *et al.* may be unique because of the karyotype of ictalurid catfish, just as the nature of the tetraploid ancestry of salmonids makes mapping of their genome unique because of phenomena such as pseudolinkage. The haploid chromosome number for channel catfish is 29. The number of linkage groups is expected to be equal to the chromosome number of 29, but initial linkage maps identify 35-40 linkage groups. The ancestral ictalurid karyotype is proposed to be 2N = 58, which is the 2N for both channel and blue catfish, with a relatively high arm number (FN), and which is probably the ancestral karyoptype for all siluriforms. FN for the ancestral ictalurid was probably greater than 80, and that for both blue and channel catfish is estimated to be 90-92. This high arm number is characteristic of ictalurids and siluriforms in general. It may not be surprising that six to 11 extra linkage groups were indicated by the microsatellite and AFLP maps, respectively, because of the large arm number, 92, of channel catfish. Channel catfish have 17 pairs of metacentric/submetacentric chromosomes and some of these AFLP and microsatellite linkage groups may represent single arms of metacentric/submetacentric chromosomes.

AFLP primer combinations affected marker distribution and number of markers in channel catfish gene mapping. Various numbers of markers were produced, depending on the primer combinations, with an average of 9.4 markers produced per primer combination. Several primer combinations produced over 20 AFLP markers; however, six primer combinations produced two or fewer AFLP markers. Additionally, the AFLP markers had an uneven distribution on the catfish gene map. A highly clustered distribution was observed for 133 AFLP markers, and these markers tended to be distributed at the end of several major linkage groups, perhaps centromeric, telomeric or both. Their distribution at the end of these linkage groups does not confirm a telomeric or centromeric location, as there were 11 more linkage groups than chromosome pairs, eventually requiring combining of some linkage groups. Therefore, many of the clustered AFLP markers could be at a position close to centromeres, or alternatively near telomeres. Additionally, several very small linkage groups were obtained, and some of these might belong to the 12 pairs of chromosomes that are subtelocentric in channel and blue catfish.

Clustering may be a characteristic of AFLP markers, as this phenomenon has been observed in other gene maps. Highly clustered AFLP markers were found in potato, barley and soybean and clustered near centromere regions in *Arabidopsis thaliana.* In the case of fish, AFLPs appeared to be primarily clustered near centromeres in rainbow trout.

Different markers appear to have different distributions in fish, as variable-number tandem repeats were frequently more telomeric and salmonid-specific small interspersed nuclear elements were intermediate in distribution compared with each other and AFLPs in the rainbow trout. In the case of zebra-fish, the distribution of the ESTs was very uniform, ranging from 73 to 201 ESTs per linkage group, indicating that there are no gene-rich or gene-poor chromosomes in zebra fish, and zebra-fish chromosomes can be identified by AT-rich repetitive sequences at the centromere and GC-rich sequences adjacent to the centromeres. This may be related to the fact that certain AFLP primers tend to amplify regions rich in specific nucleotides in catfish, and this might influence clustering and distribution patterns. This may be true for other aquatic organisms as well, and marker type definitely has an influence on the type of coverage and marker distribution pattern in the fish gene map. Preliminary results hint at consistent results among fish when using the same class of marker.

Several explanations have been put forward for the AFLP clustering in gene maps, such as the possibility that a small proportion of the clustered markers result from allelism between some AFLP bands, since AFLP are allelic markers, a reduced recombination rate around centromere regions and/or telomere regions, an actual enrichment of AFLP markers in these regions due to uneven distribution of restriction sites, the presence of highly repetitive elements within these genomic regions with great variation in both the lengths and sequences among the repetitive elements, or a combination of these explanations. The pericentromeric regions contain mainly repeated sequences of unknown functions in *Arabidopsis.* High levels of marker clustering hinder the effectiveness of AFLP markers and therefore warrant study of the nature of the genomic sequences surrounding the regions of clustering markers.

Segregation distortion is common in fish gene maps, and was observed for 101 of 607 markers in ictalurid catfish. The distorted markers appeared to be correlated with certain primer combinations. When primers *Mse*I-CAC and *Mse*I-CTG were used, large numbers of markers segregated in non-Mendelian ratios.

The distortion rate of 16% for catfish AFLP markers was lower than that found for other organisms when AFLPs were utilized to construct gene maps—65% for club-root and 54% for silkworm. Several reasons may account for the observed marker distortion in channel catfish and other aquatic organisms. Channel catfish have highly abundant Tc1-like transposable elements, and fragments amplified from such elements or other types of repetitive elements may not follow Mendelian segregation ratios. Additionally, competition among gametes for preferential fertilization, sampling in finite mapping populations or amplification of a single-sized fragment derived from several genomic regions may cause marker distortion. Segregation distortion of microsatellites in Atlantic salmon appeared to be related to preselection of eggs or embryos.

These distortions in fish gene mapping are expected, as linkage disequilibrium is caused by mixing of highly differentiated genotypes—species certainly qualify—and selection for heterozygous genotypes, which are purposely generated in mapping resource families. Perhaps linkage disequilibrium could also be caused by selection for homozygous genotypes. If selection for heterozygotes or homozygotes has occurred in these fish gene-mapping studies, the distorted markers that were unmapped because of distortion could, in fact, be important and of significance for QTL mapping, selective genotyping and MAS.

QTL MAPPING

A great acceleration in genomics and gene-mapping research for aquatic organisms has recently occurred. A large number of fish genes and regulatory sequences have been identified and isolated, the structure of the fish genome is much better understood and extensive gene maps for fish, oysters and shrimp have recently been generated. QTL mapping and MAS are becoming a reality. QTL markers for growth, feed-conversion efficiency, tolerance of bacterial disease, spawning time, embryonic developmental rates and cold tolerance have been identified in species such as channel catfish, rainbow trout and tilapias. A large number of molecular markers are needed to map QTLs and economic trait loci (ETLs) for MAS programmes and for cloning genes from various organisms for gene transfer and genetic engineering.

All types of biochemical and molecular markers are potentially useful for QTL mapping. Genes coding for isozymes and other actual genes have an inherent advantage because, of course, they are actual genes and thus have a good probability of being correlated with quantitative and qualitative traits of economic importance. Again, their disadvantage is that they are not numerous. ESTs are a solution to this problem, but much follow-up work is needed to identify polymorphism in the ESTs. The EST approach is productive for generating type I (expressed sequence) markers for gene mapping. Although many other types of molecular markers are currently available for use in genome mapping, ESTs represent actual genes. Because DNA sequences are more conserved in genes than in non-expressed sequences, comparative anchorage maps can be constructed using ESTs. Type I markers—known genes—which facilitate mapping of QTLs of aquaculture performance traits important to MAS, have been identified and are numerous.

In contrast, neutral DNA markers have the dual advantage for QTL analysis of being abundant and highly polymorphic. Microsatellites are particularly good candidates for QTL mapping and MAS as they are sometimes found within genes, they are highly polymorphic and they are inherited codominantly. Microsatellites have been identified within ESTs, and these microsatellites can be potentially useful for genomic mapping if they are polymorphic. Microsatellites were found in 4.6% of skin ESTs in channel catfish. Polymorphic microsatellites were found within dopamine receptor, profiling ribosomal protein S16, S100-calcium-binding protein A14, urokinase receptor and protein-tyrosine-phosphatase IF1 genes of channel catfish. High levels of polymorphism in the channel catfish myostatin gene should facilitate genomic mapping of this gene. Variation was found in microsatellites and SNPs in myostatin.

Theoretically, there are certain principles that can affect the reliability and applicability of QTL analysis. Care must be taken when searching for QTLs from mixtures of populations with different gene frequencies, as Hardy-Weinberg disequilibrium is not a powerful tool for detecting admixtures and, if populations are mixed, many false-positive associations can occur. Additionally, correlations between a molecular marker and disease susceptibility or resistance in a single generation of evaluation of an aquaculture species may be false because aquaculture brood stocks usually consist of relatively few groups of siblings and/or admixed populations that have different marker gene frequencies, according to Deng. He suggests that possible solutions to this QTL problem include controlled breeding and progeny testing or doubled-haploid analyses.

Novel computerized QTL mapping may be on the horizon. Grupe *et al.* have developed an accelerated QTL mapping procedure—*in silico* mapping—which can be utilized if homozygous lines, phenotypic data on the lines and a marker database are available. A computational procedure predicts the chromosomal regions that regulate the phenotypic traits, using a database of SNPs. A linkage-prediction program scans a murine SNP database and, on the basis of known inbred strain phenotypes and genotypes, predicts the chromosomal regions most likely to contribute to complex traits. This computational prediction method does not require the generation and analysis of experimental intercross progeny, but still correctly predicted the chromosomal regions identified by analysis of experimental intercross populations for multiple traits analysed. Of course, the experimental crosses had to be produced to verify that the program worked. A total of 19 of 26 experimentally verified QTL loci affecting ten traits were correctly identified.

QTLs of Aquatic Organisms

Chromosomal positions have been determined for some quantitative and performance traits in aquatic organisms. A great amount of QTL research has been conducted for sex determination and sex linkage in fish because of the great interest in producing monosex populations of certain species. Sex-linked inheritance in fish was first reported by Aida in medaka. Chromosomal positions have been identified for single genes in *Xiphophorus* livebearers, which determine age and size at sexual maturity. Waldbieser *et al.* examined 293 polymorphic microsatellite loci in channel catfish, and seven of these loci were closely linked to the sex-determining chromosome region. *Oreochromis* QTLs for sex and colour have been mapped. Young *et al.* conducted gene mapping for rainbow trout by utilizing doubled haploids -androgens. The sex-determining locus was at a distal position on one of the chromosomes. Sex-linked markers have been found for medaka.

May *et al.* backcrossed second-generation sparctics *(Salvelinus fontinalis* X *Salvelinus alpinus)* to *S. fontinalis,* allowing identification of a tight classical linkage of phenotypic sex, *Sex-1,* the primary sex-determining locus in salmonids, with *Ldh-1, Aat-5* and *Gpi-3.* The map order was centro-*mere-Ldh*-1–*(Aat-5* and *Gpi-3),* with the latter two loci being tightly linked. There was no association of phenotypic sex (presumably *Sex*-1) with these same three loci and other loci known to be linked to these loci from maps generated from splakes (*S. fontinalis* X *Salvelinus namaycush)* and cutbows *(Oncorhynchus mykiss* X *Salmo clarki).* The linkage of *Sex-1* with these loci is only observed in *S. alpinus,* indicative that *Sex-1* lies across the centromere from these three loci in *S. alpinus,* and represents a Robertsonian fusion not found in any of the other four species. Sex-determining genes and their locations vary in fish.

QTLs for growth and feed-conversion efficiency have been identified in fish and shellfish. Chromosomal positions have also been identified for single genes in *Xiphophorus* that control growth rate from birth. Hallerman *et al.* observed that several isozyme loci were correlated with growth in channel catfish. In contrast, variation at 14 allozyme loci, as measured by expected heterozygosity, was not lost after two generations of selection for weight in oysters, indicating a lack of correlation between growth and isozymes for this species. This may be a result of the vastly different reproductive strategy of oysters, discussed later. Putative linked markers to the traits of feed-conversion efficiency and growth rate have been identified for channel catfish. Tanck *et al.* utilized 11 microsatellites and found that they were correlated with mass and length in common carp. Length polymorphisms have been detected in the growth-hormone gene *(GH-2)* intron of pink salmon and could serve as QTLs, and a null allele at this microsatellite locus was responsible for an apparent deficit in heterozygotes in populations of pink salmon.

QTLs have also been identified for disease resistance in fish. In salmonids, a DNA marker linked to infectious haemato-poietic necrosis (IHN) virus disease resistance has been identified and associated with resistance to this disease in rainbow trout. Ozaki *et al.* used 51 microsatellite markers to identify several chromosome regions containing putative QTL genes that affect resistance to infectious pancreatic necrosis (IPN) in rainbow trout. Two putative QTLs affecting disease resistance were detected on chromosomes A (IPN R/S-1) and C (IPN R/S-2), respectively, suggesting that this is a polygenic trait in rainbow trout. MHC polymorphism is linked to immunity to infectious diseases in rainbow trout. QTLs for IPN virus have been obtained for rainbow trout. Putative microsatellite markers are linked to resistance to the bacterium *Edwardsiella ictaluri* in channel catfish. Tanck *et al.* utilized 11 microsatellites and found that they were correlated with stress-related plasma cortisol levels and basal plasma glucose levels in common carp.

QTLs for fitness traits and survival have been identified in fish and shellfish. A microsatellite accounted for 7.5% of the variance in thermal tolerance in unselected populations of rainbow trout. Negative QTLs can also exist. Oysters and elm trees rely on sexual recombination and high fecundities, with millions of highly varied offspring, to respond adaptively to environments that are highly heterogeneous in space and time, and usually exhibit strong heterosis in nature, with homozygous offspring at marker loci expiring quickly as a cohort of offspring ages. This can result in unusual, non-Mendelian inheritance of allozyme and microsatellite marker loci. Heterosis and segregation distortion are due to linkage between the (neutral) markers and deleterious recessive alleles at nearby loci in Pacific oysters. The heterosis in Pacific oysters was due to linkage, not to the intrinsically higher fitness of shellfish that are heterozygous for the markers, and the non-Mendelian inheritance of markers in the oyster was due to the selection against deleterious homozygotes at linked loci. Oysters carry a high load of deleterious mutations, but the strategy of producing millions of recombinant offspring allows them to thrive. Similar results have been obtained in *Drosophila*, which demonstrated that evolution proceeds in sexual organisms because recombination allows favourable genes to overcome deleterious genetic backgrounds. These factors affect the ability to identify QTLs and, in some cases, identification of negative QTLs to select against may be as valuable as identifying positive QTLs to select for.

Strong relationships also exist between DNA markers and reproduction in fish. Fishback *et al.* examined the maturation of a domestic strain of rainbow trout for which the spawning season had been expanded from 2 weeks to 8 months through selection. The spawning time for the majority of the females, but not the males, could be predicted based on genotypes from 14 microsatellite loci. The distributions of the neutral microsatellites diverged along with the selective divergence of spawning date, and the changes in spawning date were correlated with changes in the frequencies of QTL markers.

Marker-assisted Selection

Theoretically, MAS has the potential to greatly accelerate genetic improvement in livestock, aquatic organisms and plants. Highly saturated genetic maps have been generated for rice, maize, wheat, tomato, cotton, soybean, cattle, pigs and sheep as well as aquatic organisms, providing the genetic framework and tools for developing MAS programmes. MAS potentially enables improvement in economically important traits and may provide a powerful alternative for improving traits that are difficult to breed for, such as carcass yield, disease resistance, feed-conversion efficiency and sex-limited traits, compared with traditional approaches, such as family and indirect selection. MAS may be particularly useful for low-heritability traits and those complicated by dominance effects. Theoretical calculations predict that MAS would increase the rate of genetic gain by 25-50% over traditional animal-breeding programmes. Traditional selective breeding in cattle, pigs and sheep results in a genetic progress per year of approximately 1% for several traits. MAS may be less dramatic for fish than for livestock in comparison with selection for improving growth, since typical genetic gain for fish growth rates are 6-14% per generation, which is equivalent to 2-14% per year, with an average of about 3-4% per year.

To commercialize MAS technology, QTLs are located and their effects on the phenotype measured. Then the markers are evaluated in commercial populations. Lastly, the markers are combined with phenotypic and pedigree information in genetic evaluation for predicting the genetic merit of individuals within the population to allow actual MAS.

Marker-assisted Selection in Fisheries

Only a few examples of actual MAS exist for fish. The growth rate of rainbow trout was increased by 26% based on selection for an mtDNA marker, but this method was strain-specific because the relative performance of fish with the specific haplotype was consistent across males within strains but not across strains. In contrast, six generations of traditional selection increased body weight by 30% in rainbow trout. MAS improved feed-conversion efficiency by 11% for aquaculture species, while traditional selection improved feed-conversion efficiency by 4.3%. In rainbow trout, 25% of progeny exhibited a high degree of upper-temperature tolerance after MAS for heat tolerance. It appears that MAS has the potential to accelerate genetic improvement of aquaculture species.

REFERENCES

Dunham, R.A. and Liu, Z. (2002) Gene mapping, isolation and genetic improvement in catfish. In: Shimizu, N., Aoki, T., Hirono I. and Takashima, F. (eds) *Aquatic Genomics: Steps Toward a Great Future.* Springer-Verlag, New York, pp. 45–60.

Liu, Z.J. and Dunham, R. (1998a) *Genetic Linkage and QTL Mapping of Ictalurid Catfish.* Circular Bulletin 321, Alabama Agricultural Experiment Station, 19 pp.

Morizot, D.C. and Siciliano, J.J. (1984) Gene mapping in fishes and other vertebrates. In: *Evolutionary Genetics of Fishes.* Monograms in Evolutionary Biology, pp. 173–233.

Whitehead, P.K. (1994) Mapping and electroporation of mitochondrial DNA from three species of ictalurid catfish. Doctoral dissertation, Auburn University, Auburn, Alabama, USA.

9

Environmental Impacts of Fish Biotechnology

Commercialization of transgenic aquatic organisms on a large scale may have a variety of ecological implications. Eventual escape of transgenic aquatic organisms from confinement will occur from a commercial facility and the range of receiving ecosystems is broader.

Much concern exists concerning the potential food safety ethics of utilization and ecological impacts of transgenic fish. In addition to benefits, aquatic genetically modified organisms (GMOs) may also pose environmental and food-safety hazards. Potential ecological hazards include adverse interactions with a range of species with which a GMO interacts in the accessible ecosystem, and genetic hazards to conspecific natural populations. Ecological hazards include the possibility of increased predation or competition, colonization by GMOs in ecosystems outside the native range of the species and, possibly, alteration of population or community dynamics due to the activities of GMOs.

Fertile GMOs could interbreed with natural populations, and any genetic or evolutionary impacts, positive, negative or neutral, would depend on the fitness of the new genotypes in the wild. Risk would exist when fitness relative to the wild type is high, and also when maladaptive traits and genes might be introduced into native populations, although, logically, these would be selected against.

The ecological, genetic and evolutionary impacts of GMOs in the range of relevant aquatic and marine systems, and the ecological and genetic risk pathways and end-points posed by commercial-scale application of aquaculture biotechnology, need to be thoroughly studied. Interdisciplinary approaches to environmental risk assessment and monitoring will be needed to totally understand the effects of transgenic aquatic organisms in the environment. Risk will need to be evaluated on a temporal basis. Rapidly occurring phenomena, such as large-scale escapement, needs to be compared with slowly occurring phenomena, such as the adaptive evolution of transgenic aquatic organisms in the environment. Spatial connections among aquatic ecosystems will also need consideration.

Concerns about hazards posed by aquatic GMOs have been inferred on the basis of ecological principles; however, the totality of the impact and how it might occur are more complex than what is presented in these papers based on principles. Existing experimental evidence indicates that most transgenics pose little ecological risk and a subset of transgenics may pose a potential risk, although these data were derived from models in highly artificial situations. Questions concerning these issues are beginning to be answered, and it has been rumoured that commercialization of transgenic fish has taken place in China and Cuba. However, government officials of these two countries indicate that commercialization has not occurred, but that potential commercialization is being evaluated. Brood stock for potential commercialization of transgenic salmon were once present in New Zealand, but are now probably in Chile. Transgenic salmon will probably be approved for human consumption in the USA in 2004, but approval for the growth of these fish in the USA will not be granted at the same time. A non-food fish, transgenic zebra fish containing fluorescent pigment genes, has been commercialized in both Asia and the USA. However, in locations such as Europe and Japan, conservative approaches to the development of transgenic fish will prevail politically for many more years. Because of these concerns, transgenic fish will probably be utilized commercially to a greater extent in developing countries than in developed countries in the short term.

Transgenic fish, assuming they are derived from domestic strains, may not have any more genetic impact on natural populations than domestic conspecifics. However, genetic modifications that would allow expansion of a species' range—essentially the development of an exotic species—would probably have the greatest ecological impact. For instance, the development of a cold-resistant tilapia or a cold-resistant salmon with antifreeze protein would allow these fish to expand their geographical range. As an exotic species, they would interact with local biota and have the potential for ecological impact, as alterations in species composition is considered detrimental.

The majority of introductions of exotic fish are unsuccessful. Successful introductions are more likely to occur in temperate rather than tropical habitats, in environmentally stressed fish communities, in simple rather than diverse fish communities and in environmentally stressed habitats. About 10% of attempted introductions are successful and, of these, 10-20% result in species introductions that are considered to have adverse ecological effects. However, when they occur, the adverse impacts can be severe. One example is the introduction of Nile perch in Lake Victoria resulting in the extinction of several cichlids. The introduction of predatory species of transgenics may have the potential for larger effects than the introduction of prey species.

Because of these concerns about transgenic aquatic organisms, research on food safety and potential environmental impact, including the measurement of fitness traits, such as predator avoidance, foraging ability, swimming ability and reproduction, is needed to allow educated decisions on the risk of utilizing specific transgenic fish. These data will be necessary for the application of transgenic fish in North America and Europe.

The impact of domestic aquacultured organisms, interspecific hybrids, polyploids and genetically engineered fish on the genetic variation of conspecifics, population numbers and performance of conspecifics and the ecosystem in general, is currently being questioned, debated and researched. Data are building up concerning the interactions between domestic and wild populations and the fitness of genetically enhanced aquatic organisms such as to allow policy development and management application to be based solely on scientific fact and principle.

Certain triploid aquatic organisms may also pose risk under particular circumstances. Triploid male grass carp and salmonids undergo sexual maturation and may seek matings, which would result in loss of the resulting broods, posing demographic risk to the small natural populations they may encounter. Triploid Pacific oysters, *Crassostrea gigas*, can exhibit reversion to diploidy, although it is not yet known whether this restores fertility. The effectiveness of triploidy as a means of limiting risks associated with introductions of non-native species, genotypes or transgenics needs further evaluation.

Assuming that environmental risks associated with transgenic fish exist, benefit-risk analysis will be required. Ecologically, the primary concerns regarding the utilization of transgenic fish are loss of genetic diversity, loss of biodiversity and changes in the relative abundance of species upon release or escape followed by establishment of transgenic fish in the natural environment. Conversely, the utilization of transgenic fish in aquaculture could actually enhance genetic diversity and biodiversity by increasing food production and production efficiency, thus relieving pressure on commercial harvests of natural populations and decreasing pressure on land and water use for agriculture and aquaculture.

In the past, society decided to increase population size, expand agriculture, exploit natural populations and dam rivers to generate electricity, feed people and increase the quality of life at the expense of biodiversity. Aquaculture, if properly implemented, and genetic improvement have the potential to increase production and fish availability and to decrease pressure on wild stocks, thus preserving natural genetic diversity. Captive stocks, if properly managed, could also be utilized to preserve genetic diversity. Cost-benefit analysis will be necessary to assess whether or not transgenic fish application is warranted.

The extent of phenotypic change from the introduction of a fusion gene could be analogous to the development of an exotic species, a select line or a domestic strain, depending upon the magnitude of the phenotypic change. Transgenic fish could express phenotypes that are analogous to the formation of a new exotic species. Approximately 11% of introductions of exotic species actually become established and of these about 10% have negative ecological impacts.

The genetic impact of genetically improved aquaculture fish could have neutral, positive or negative effects on wild populations, in the short term or long term. Transgenic fish could escape and the trans-gene become part of the gene pool. This could add genetic diversity to the population, lower or raise fitness or have no phenotypic or ecological effect. If there is a lowering of fitness, it should be temporary as the trans-gene should be selected against. The mixing of the gene pools might enhance genetic resources, increase genetic variation or result in heterosis, all potentially positive results. The wild fish may outcompete and eliminate the domestic fish or the domestic fish may have no long-term impact on the performance of the population—neutral or non-effects. Negative impacts could result from outbreeding depression, which would theoretically be temporary, or from the elimination of wild genotypes through competition.

Although the concern that the accidental or intentional release of domestic and genetically improved fish will have a damaging effect on native gene pools is legitimate and requires careful scrutiny, the available data indicate that the potential damage of domestic fish to native conspecific gene pools is quite small. First, it is even difficult to change the genetic make-up of an established native gene pool with the intentional stocking of wild conspecifics from another watershed. Secondly, wild fish almost always outcompete their domestic genotypes in a natural setting.

Most data indicate that wild fish are more competitive than domestic fish, resulting in the elimination of the domestic fish and their potential positive or negative impacts. However, recent evidence from salmonid research indicates that there are situations where domestic fish can have a genetic impact on wild populations. When repeated large-scale escapes of domestic fish occur, genetic impact can occur just from the swamping effect of sheer force of numbers. Transgenic fish could make an impact in this scenario of large-scale escape, but again the consequences should not vary much from that of fish genetically altered by other means.

Most types of transgenic fish, including those that are growth hormone (GH)-gene transformants, are more analogous to a selected line or domestic strain. The change in phenotype is similar to what would be observed in or what would be the goal of strain selection, selection, intraspecific crossbreeding, interspecific hybridization, sex reversal or gynogenesis. If a fourfold increase in growth is possible through traditional breeding (and such gains are possible), ecological impacts would be the same regardless of the mechanism of phenotypic alteration—traditional or biotechnological.

Transgenic fish should be analogous to select lines and domestic strains in a second manner. Transgenic fish will probably be generated from select lines and domestic strains since these fish are generally more suited for aquaculture and already have increased performance in the aquaculture environment.

Obviously the primary benefits of transgenic fish would be increased aquaculture production and profitability. However, other potential benefits exist in addition to those mentioned earlier.

Escaped transgenic fish could add genetic diversity to populations. This would be artificially induced genetic diversity, which some sectors of society would value and to which others would be opposed. This artificial genetic diversity could actually increase fitness in some endangered populations or species and make such genetic units more viable. For example, natural and human-induced factors have apparently reduced genetic variation in the cheetah to the point where reduced reproductive performance threatens their existence. Gene transfer would be an option to restore reproductive performance and save this species.

Evidence exists that when humans began exploiting fish populations much more efficiently and intensively in the last 200 years, certain traits, such as size, were genetically selected against. It is likely that have permanently eliminated important growth alleles and perhaps alleles for other traits from some fish species. Gene transfer is an option to restore phenotypes that have been artificially eliminated.

The escaped transgenic fish could replace the natural population. Depending upon the existence or absence of this genotype, genetic diversity would be lost. The long-term survival of that species or population at the location could be enhanced, decreased or unchanged. Environmental risk data to date, however, indicate that the above scenario, replacement of the natural population, is unlikely.

Transgenic fish could become established and hybridize with other species, spreading the transgene to other species. This scenario is unlikely since reproductive isolating mechanisms usually restrict permanent gene flow between species of fish.

All available data indicate that transgenic fish are less fit than non-transgenic fish and would probably have little, if any, environmental impact. Additionally, domesticated transgenic fish would be expected to have less environmental risk than wild transgenic fish, based on the discussion above. However, the greatest environmental risk that a transgenic fish would have is when the gene insert would allow the transgenic genotype to expand its geographical range, essentially becoming equivalent to an exotic species. About 1% of such releases of exotics result in adverse environmental consequences.

Altering temperature or salinity tolerance would be analogous to the development of an exotic species since this would allow the expansion of a species outside its natural range. This type of transgenic research and application should be avoided. Antifreeze-protein genes from winter flounder have been introduced into Atlantic salmon in an attempt to increase their cold tolerance. If this research were successful, a real possibility of environmental impact exists. Similarly, if tilapia were made more cold-tolerant, a strong possibility of detrimental environmental impact exists. Sterilization could reduce risk, but genetic means of sterilization, such as triploidy decrease performance. Additionally, fertile brood stock are necessary, so risk is minimized but not eliminated. Transgenic sterilization, to be discussed later, is potentially a much better option than triploidy.

Currently, it is common practice in Asia to introduce exotic species to address shortcomings in the aquaculture performance of native species. Again, the introduction of exotic species has the greatest potential to adversely affect biodiversity and, consequently, genetic diversity. The utilization of transgenic fish derived from the indigenous aquaculture species is more likely to be an environmentally safe means of addressing the perceived aquaculture shortcomings of native species and is less likely to decrease biodiversity and genetic diversity compared with the continued practice of exotic-species introduction.

Interspecific hybrids pose similar environmental risks to those of transgenic aquatic organisms. In some cases, usually where parent species have limited geographical distributions, but very rarely, hybrids introgress permanently with parent species in a natural setting. It appears to be a greater problem in maintaining aquaculture populations pure for hybridization programmes.

Environmental Risks of Transgenic Fish

Efforts should be organized to evaluate the potential environmental risk of transgenic fish. Reproductive performance, foraging ability, swimming ability and predator avoidance are the key factors determining fitness of transgenic fish and should be a standard measurement prior to commercial application. Most ecological data on transgenic fish gathered to date indicate a low probability of environmental impact. Extremely fast-growing salmon and loach have low fitness and die.

Several models have been developed that estimate and indicate genetic risk of transgenic fish. Muir and Howard evaluated a model and described the Trojan gene effect: the extinction of a population due to mating preferences for large transgenic males with reduced fitness. Therefore, reduced fitness as well as increased fitness has potential adverse ecological effects. This modelling was based on experimental results of medaka in aquariums.

Hedrick developed a deterministic model indicating that, if a transgene has a male-mating advantage and a general viability disadvantage, analogous to the Trojan gene effect of Muir and Howard, then in 66.7% of the possible combinations for possible mating and viability parameters for its invasion in a natural population, the transgene increases in frequency and, for 50% of the combinations—the possible combinations of the possible mating and viability parameters—the transgene goes to fixation. The increase in the frequency of the transgene reduces the viability of the natural population, increasing the probability of extinction of the natural population.

In another modelling exercise, Muir and Howard again conclude that a transgene is able to spread to a wild population even if the gene markedly reduces a component of fitness, based on data from a laboratory population of medaka harbouring a regulatory sequence from salmon fused to the coding sequence for human GH. The juvenile survival of transgenics was reduced in the laboratory but the growth rate increased, resulting in changes in the development rate and size-dependent female fecundity. The important factors in the model were the probabilities of the various genotypes mating, the number of eggs produced by each female genotype, the probability that the eggs would be fertilized by the sperm of each male genotype (male fertility), the probability that an embryo would be a specific genotype given its parental genotypes, the probability that the fry would survive and parental survival.

Muir and Howard's interpretation was that transgenes would increase in populations despite high juvenile viability costs if transgenes also had sufficiently high positive effects on other fitness traits. Sensitivity analyses indicated that transgene effects on age at sexual maturity should have the greatest impact on transgene allele frequency. Juvenile viability had the second greatest impact. A defect in the simulation was the fact that the effect of predation in the wild could not be included in the model, biasing viability estimates. Although these modelling experiments based on laboratory data on small model species illustrate the potential risk of transgenic fish, some weaknesses in the analysis exist. The environment was artificial, the mating preference does not exist for many fish, including catfish, the data were not put into the model to account for genotype-environment interactions, which are likely, predation is absent, as Muir and Howard indicate, and the overall performance of the fish is not accounted for.

Body size does not necessarily result in mating advantages. Rakitin *et al.* utilized allozymes and minisatellites to determine that male size, condition factor and total or relative body-weight loss over the season were not correlated with the estimated proportion of larvae sired by each Atlantic cod male during the spawning season. Similar results were observed in salmon. However, Atlantic cod male reproductive success was affected by female size, with males larger (> 25% total length) than females siring a smaller proportion of larvae. In this case, large size was reproductively disadvantageous.

Although there may be cases where size increases reproductive fitness of both sexes, cultured transgenic fish may not be allowed to grow large enough to have a mating advantage as escapees, and data on transgenics of larger aquaculture species have not shown any reproductive advantage for the transgenics. Fast-growing transgenic tilapia have reduced sperm production. Transgenic channel catfish and common carp have a similar reproduction and rate of sexual maturity compared with controls. The spawning success of transgenic channel catfish and controls appeared similar. When the two genotypes were given a choice in a mixed pond, the mating was random and the spawning ability of transgenic and control channel catfish was equal.

Genotype-environment interactions are important and occur for the growth of transgenic channel catfish. Transgenic channel catfish containing salmonid GH genes grew 33% faster than normal channel catfish in aquaculture conditions with supplemental feeding. However, there was no significant difference in growth performance between transgenic and non-transgenic channel catfish in ponds without supplemental feeding, indicating equal foraging abilities and the inability of transgenic catfish to exhibit their growth potential with limited feed. The foraging ability of transgenic and control catfish is similar under these conditions of competition and natural food sources and growth is no different between transgenic and control catfish in these more natural conditions. When grown under natural conditions where food is limiting, the transgenic channel catfish has a slightly lower survival than controls and grows at the same rate as non-transgenic controls. As is the case in most genetic improvement programmes, genetically altered fish need adequate food to express their potential.

The faster-growing transgenic fish could have impaired swimming, leading to predator vulnerability, problems in capturing prey, reduced mating ability for some species and reduction in competitiveness for any trait requiring speed. Selection for swimming ability may be one of the primary mechanisms limiting the genetic increase in size of fish and preventing fish from evolving to larger and larger sizes.

Silversides, *Menidia menidia,* from Nova Scotia ate more food, had more efficient feed conversion and grew faster than a population from South Carolina. However, the maximum prolonged and short-term swimming speeds of Nova Scotia strain were lower than those of the South Carolina strain, and the swimming speeds of fast-growing phenotypes/genotypes were lower than those of slow-growing phenotypes/genotypes within each strain. Slow swimming speed has a fitness cost: vulnerability to predation. The Nova Scotia strain was more vulnerable to predation than the South Carolina strain, and predation increased with growth rate and feeding rate both within and between strains. Maximizing energy intake and growth rate engenders fitness costs in the form of increased vulnerability to predation.

In fact, initial experiments all indicate that the predator avoidance of transgenic fish is inferior compared with controls, as would be predicted by the experiment with the silver-sides.

Predator avoidance was slightly better for non-transgenic catfish fry and fingerlings when exposed to largemouth bass, *Micropterus salmoides,* and green sunfish, *Lepomis cyanellus,* than for transgenic channel catfish. Data on salmon also indicate that they probably have reduced fitness for non-aquaculture, the natural environment. GH-transgenic salmon have an increased need for dissolved oxygen, a reduced swimming ability and a lack of fear of natural predators.

On an absolute speed basis, transgenic coho salmon swam no faster at their critical swimming speed than smaller non-trans-genic controls, and much more slowly than older non-transgenic controls of the same size. Again, as was found with the silversides, a marked trade-off was observed between growth rate and swimming performance. Farrell *et al.* hypothesized that the decreased swimming ability may be a result of some physiological change due to the hyperlevels of GH excretion. However, Ostenfeld *et al.* offer an alternative explanation. Coho salmon containing the *Oncorhynchus* metallothionein GH1 plasmid (pOnMTGH1) had an altered body contour and centroid size and enhanced caudal peduncle and abdominal regions compared with controls. The most prominent alterations were the change in the syncranium and that the head of the transgenics was less elliptical. The opercular series were shifted, with an enlargement of the branchiostegal and augmentation of both the opercular and the cleitrum regions. The overall body shape is less fusiform for the transgenic coho salmon. Therefore, the decrease in swimming ability may be a result of loss of hydrodynamics and increased drag coefficients caused by the altered body shape. This change in body shape might also alter leverage or efficiency of the muscle movements for swimming. The inferior swimming ability of the transgenic salmon should cause them to have inferior predator avoidance, inferior ability to capture food and inferior ability to migrate to reach the sea or return to reproduce.

Transgenic fish could be more competitive in seeking feed. Devlin *et al.* examined the ability of F_1 coho salmon (250 g) containing a sockeye MT-B promoter fused to the type 1 growth hormone gene-coding region to compete for food through higher feeding motivation. The consumption of contested food pellets was determined by matching pairs of one sibling control or by size-matching pairs of one control (1 year older non-transgenic coho salmon) and one GH-transgenic coho salmon, and then determining which fish captured the first three pellets presented one at a time at each feeding trial. The transgenic coho salmon consumed 2.5 times more contested pellets than the sibling controls and the transgenic fish consumed 2.9 times more pellets than the non-transgenic size controls, indicating the high feeding motivation of the transgenic fish, throughout the feeding trials. The shortcomings are that this is a highly artificial environment and a food type that will not be encountered under natural conditions.

F_2 transgenic Atlantic salmon contained a salmon GH gene that was continuously expressed in the liver, enhancing growth 2.62- to 2.85-fold over the size range 8-55 g and improving feed-conversion efficiency by 10%. These transgenic fish had higher metabolic rates, but they consumed 42% less total oxygen between hatching and smolt size and, when starved, the rate of oxygen consumption declined more rapidly in the transgenic Atlantic salmon. The starved transgenic Atlantic salmon also lost protein, dry matter, lipid and energy more quickly than controls. The persistence of transgenic Atlantic salmon in maintaining a higher metabolic rate, combined with their lower initial endogenous energy reserves, suggests that the likelihood of growth-enhanced transgenic salmon achieving maximum growth or even surviving outside intensive culture conditions may be lower than that of non-transgenic salmon. These hypotheses are consistent with the data of Dunham *et al.* with GH-transgenic catfish, which did not come near to the dramatic phenotypic alterations observed in GH-transgenic salmon. These transgenic catfish grew at the same rate as controls under natural conditions, perhaps due to lack of food, higher metabolism coupled with lack of food or inferior swimming ability to allow capture of prey. These fish also exhibited higher mortality under the natural conditions—again, possibly being related to the lack of food and inability to capture food, coupled with higher metabolism and more rapid loss of nutrient stores, as well as possible differential mortality due to predation by aquatic insects and the potential slower swimming of the transgenics.

All transgenic fish evaluated to date have fitness traits that are either the same or weaker compared with controls. The increased vulnerability to predators, lack of increased growth when foraging and unchanged spawning percentage of these transgenic fish examples indicate that some transgenic fish may not compete well under natural conditions or cause major ecological or environmental damage. Although transgenic fish may be released to nature by accident, ecological effects should be unlikely because of these examples of reduced fitness. However, implementation of physical and biological (sterilization) containment methods may reduce further potential interaction between transgenic and wild fish populations.

Issues in Genetic Conservation

The preservation of genetic diversity is a common goal for both aquaculture breeders and managers of natural populations. Wild populations and their genes represent a living gene bank that is needed for future resources for genetic improvement. Therefore, transgenic fish research as well as aquaculture genetics research should be conducted in a manner that minimizes genetic impact on natural populations for sound ecological reasons, as well as protecting future resources for exploitation, except in situations where genetic impact on the natural population is desirable.

Production of transgenic fish and aquatic invertebrates is an extremely promising approach to enhancing global food security and efficiency by developing high-performance aquatic organisms. Transgenic fish may actually provide better protection of natural genetic resources by relieving pressure on natural exploitation and decreasing the need for the destruction of habitat for increased food production. Early evidence indicates that high-performance transgenic fish may actually have low fitness, decreasing the likelihood of their establishment in the wild and of associated potential impacts. Transgenic fish and aquatic invertebrate research is now conducted in many countries. Transgenic fish development is inevitable. The organized development of these programmes would help ensure that environmental risk and fitness traits, as well as food-safety issues, are addressed. The establishment of collaborative networks to develop protocols for and to conduct sound and safe research on transgenic aquatic organisms would help to ensure that the benefits rather than the detriments are the product of aquaculture gene-transfer research throughout the world.

Genetic Sterilization

Genetic enhancement of farmed fish has advanced to the point that it is now having an impact on aquaculture worldwide; however, potential maximum improvement in overall performance is not close to being achieved. Examples exist that indicate that greater genetic gain can be obtained in one or more traits by simultaneously utilizing more than one genetic strategy. Overall performance can probably be maximized by combining the advantages of selection, intraspecific crossbreeding, interspecific hybridization, polyploidy and genetic engineering. There are environmental concerns regarding the application of domestic, hybrid and transgenic aquatic organisms.

Additionally, the aquaculture industry's ability to capitalize on market potential is hampered by community and scientific concern over escapees from aquaculture facilities and their impact on aquatic ecosystems. Debate over the costs/benefits of these issues has divided communities and resulted in the development of strict environmental policies and legislation that often preclude growers from producing the aquaculture product of choice. Utilization of exotic species is especially unacceptable for many government and stakeholder groups, because of the perception and sometimes reality of the high risk of escapees establishing feral populations and having a negative impact on local ecosystems.

Aquacultured organisms, such as Pacific oysters in Australia and Atlantic salmon in British Columbia, and recreational/commercial species, such as Nile perch, have established destructive feral populations, creating environmental problems. Additionally, stocked domestic and wild conspecifics have the potential to alter the allele frequencies of established native populations, limiting management options for natural-resources agencies. Concern about these potential environmental and genetic effects has led to restrictions on industry development at some locations. Concern over these issues is likely to grow as demand for genetically improved stock escalates to fulfil production requirements.

One option is physical containment, but in reality protection of native stocks from escapees of aquaculture production facilities by the use of physical containment cannot be guaranteed in most cases. The ultimate safeguard would be a mechanism that prevented breeding in the wild of domestic, exotic, highly selected or transgenic stocks. Such a mechanism would prevent cultured aquatic organisms from establishing feral populations, preventing genetic pollution of local strains from domestic conspecifics and protecting the intellectual property invested in highly selected lines or genetically enhanced populations.

Combining variations of chromosome manipulation, monosex and transgenic techniques may produce sterile individuals of only one sex. For example, manipulating gonadotrophin-releasing hormone (GnRH) gene expression may cause sterility, which could be accomplished, theoretically, through antisense RNA constructs, ribozyme approaches or gene knockout. Sterilization with polyploidy, hybridization, transgenesis or combinations of these is the ultimate method to diminish these concerns and environmental risk.

However, in many cases, hybridization reduces, but does not eliminate, reproduction, so it has limitations, including the fact that fertile parental stocks must be maintained, which could escape. A better sterilization procedure is the induction of triploidy. However, the disadvantages are that triploidy also requires the existence of completely fertile brood stock, and triploidy can have adverse effects on performance, at least partially negating the genetic gain from the primary enhancement programme. Additionally, triploidy is not feasible or commercially feasible for some species, batch-to-batch variation does not guarantee that all individuals produced are triploid and triploid oysters and some triploid fish have the ability to produce small numbers of diploid progeny. Therefore, triploidy is not 100% effective for some species and triploidy would reduce the rate at which feral populations become established, but would not prevent them.

Monosex approaches could be successful for some, but not all, species. However, they would only be effective for application of exotic species, not conspecifics. Additionally, in the case of applying this technology for exotic species, it is only effective if the exotic species is not already present and, in the event of escape of the monosex/mono-genetic organisms of both sexes, short-term impacts are possible for two generations until both sexes die out. The transgenic approach is the ultimate approach.

The Commonwealth Scientific Industrial Research Organisation (CSIRO), Australia, has initiated research on an untested transgenic sterilization approach, utilizing the insertion of gene constructs that reversibly interrupt development, resulting in functional sterility. This technique by itself or in combination with others is the ideal solution for inducing sterility and protecting the environment. Theoretically, aquatic organisms would only be able to complete their life cycle under culture conditions, and escapees from captivity would be unable to breed or produce viable offspring. Progeny from fish that escape die without the intervention of humans, effectively meaning that the escaped parents are sterile. In the hatchery, a simple repressor compound is applied at a particular embryonic stage, allowing the aquatic organism to live and eventually breed. If successful, this technique has profound implications for preventing gene introgression from domestic translocations and escapes into wild gene pools, for importing exotics for culture without establishment or long-term effects on biodiverstiy and for controlling nuisance species via the sterile screwfly approach.

This technology is related to a procedure to sterilize genetically modified plants: the terminator gene or technology protection system (TPS) developed by the Delta and Pine Land Company (D&PL). The method stops the seeds of certain plants from germinating, and utilizes a transiently active promoter operably linked to a toxic gene, but separated from the toxic gene by a blocking sequence that prevents the lethal gene's expression. A second gene encodes a recombinase, which, upon expression, excises the blocker sequence, and a third gene encodes a tetracycline-control-lable repressor of the recombinase.

Plants transgenic for all three genes grow normally and are fertile. Seeds sold to farmers are treated with tetracycline, which activates expression of the recombinase, which then excises the blocker sequence. These seeds germinate, but once the blocker sequence is removed, expression of the toxin gene occurs, causing these plants to produce seeds that are sterile. Theoretically, this technology works, but its existence has not been verified. This technology may be difficult to duplicate exactly in aquatic organisms, as few recombinases have been identified that will function in animals, and those that have been identified, Cre and Flp recombinase, function in only a limited number of species. This technology has met great opposition because it forces the farmer to repeatedly return to the vendor for seeds, and the controversy has been so great that the technology has been temporarily abandoned for plants. However, because of the tremendous potential mobility of and the

concern for transgenic aquatic organisms and their perceived potential impact, such technology for aquatic organisms should be welcomed, in contrast to the situation with less mobile plants. Other strategies, such as trait-specific genetic use restriction technology (T-Gurt), have also been suggested to provide intellectual-property protection, but these technologies do not prevent gene transfer to wild stocks.

In the CSIRO approach, genes have been identified that are crucial for and activated only during embryonic development and/or gametogenesis. DNA constructs have been made containing a blocker to development or gametogenesis and a genetic switch to control its function. Grewe *et al.* have developed promoters, blockers and repressors for zebra fish, *Brachydanio rerio,* and Pacific oyster. Usually native promoters and blocking sequences were utilized for each species. Sterile feral gene constructs developed from these components prevent production of functional gametes or cause mortality in offspring produced by escapees mating outside a controlled hatchery environment. Components of the sterile feral construct are fused so that a species-specific promoter is coupled to a repressible element, which in turn drives expression of a blocker gene. In captivity, the blocker can be inactivated by triggering the promoter to allow production of the repressor protein with materials such as zinc or an antibiotic, via diet or in soluble form, so that fertilization can occur or embryos can complete development. In the wild, where the compounds to activate the repressor are not available, the blocker would remain active and disrupt the function of the critical gene, either preventing fertility or causing lethality in embryos. The promoter that drives expression of the blocker gene has a narrow spatial and temporal window of activity. Thus, addition of a specific repressor molecule to the food or water in the hatchery is only required for a brief period to repress transcription and subsequent knockout in the resulting offspring. Outside this temporal window, even in the absence the repressor molecule, the promoter is inactive and the blocker gene is not transcribed. This permits hatchery-reared off spring to survive and remain free once placed into the farm environment for grow-out. However, the promoter is expressed in the absence of the repressor molecule in any offspring that are produced outside the hatchery conditions and they die. The active promoter transcribes the blocker sequence, which leads to disruption of critical gene function and eventual mortality. The blocker gene functions as a dominant allele and thus escapees cannot produce viable offspring even if they interbreed with wild-type fish.

The same approach as above can be used to block gametogenesis. Without delivery of a repressor activator, gamete production does not occur and escapees cannot breed. With the delivery of the repressor via feed, injection or implant, gamete production is allowed, but the offspring of escapees are sterile as no repressor is available in the natural environment. Again, the blocker must act as a dominant allele so that the offspring of transgenic aquatic organisms mated with wild types are sterile.

One option for the blocker mechanism for embryogenesis is disruption of normal cellular activity to cause embryonic mortality via production of a cytotoxic protein. However, production of aquatic animals possessing a toxic gene, although biologically safe, would probably not be accepted by consumers.

An approach to achieving embryonic gene disruption, or knockout, is the use of mRNA specifically targeted to interference of gene function. Mechanisms used to achieve mRNA knockout, including the expression of ribozymes, antisense mRNA and double-stranded mRNA (dsRNA), have succeeded in several organisms. Because of the high specificity of mRNA targeting, the targets in aquaculture species are unlikely to have any close homologies in humans, which eliminates this as an issue should it ever become of concern. Currently, there are efforts to trans-genically sterilize tilapia by preventing expression of GnRH and luteinizing hormone via knockout, antisense and ribozymes, and there is some preliminary evidence of reduced fertility in these fish.

Repression of the knockout function is achieved in zebra fish by coupling a developmental stage-specific promoter to components derived from the commercially available Tet-Off controllable expression system, marketed by CLONTECH. The Tet-Off system is also functional in oyster primary cell cultures, and has also proved effective at repressing the action of a transgene system in *Drosophila*. Tetratracycline or doxycy-cline is used as the repressor molecule, and both can be easily administered either in hatchery water to allow embryonic development or in food or by injection to allow gametogenesis in brood stock. Tetracycline is a routinely used antibiotic in fish and shellfish culture and should therefore be acceptable from a food-safety perspective.

Grewe *et al.* describe the sterile feral construct as follows: the sterile feral construct accomplishes repression of the blocker gene by tetracycline or doxycycline acting upon the regulatory protein (trans-activator protein (tTA)), a fusion of TetR and VP16 as derived from the pTet-Off regulatory plasmid. A zebra-fish promoter drives expression of the tTA, and tTA regulates the tetracycline (Tet)-responsive human cytomegalovirus promoter (*P*hCMV*-1), which controls expression of the blocker gene. *P*hCMV*-1 contains the tetracycline-responsive element (TRE), which consists of seven copies of the tet operator sequence *(tetO)* located just upstream of the minimal cytomegalovirus (CMV) promoter (*P*minCMV). *P*hCMV*-1 does not function in the absence of the binding of tTA to the *tetO*. The tetracycline-sensi-tive element is described by Gossen and Bujard for Tet-Off and Tet-On is described by Gossen *et al.* and Kistner *et al.* In the Tet-Off system, the addition of tetracycline (Tet) or doxycycline (Dox), a Tet derivative, prevents the binding of tTA to the set-responsive element. Expression of the blocker gene is controlled by TRE, and is repressed until tetracycline is removed from the incubation water. In the absence of Tc or Dox, the blocker gene is transcribed as long as the zebra-fish promoter continues to express tTA, thus killing the embryo.

The lac operon system may provide another transgenic mechanism for sterilization. The essential parts of the *Escherichia coli* repressor system were altered to function at high efficiency in mice, and then transferred into mice to control the production of tyrosinase. When fed a standard laboratory diet, the transgenic mice were albino, as the repressor protein of the trans-gene blocks the operation of the mouse tyrosinase gene. When a lactose analogue is added to the diet, the repressor protein changes its shape and disconnects from the DNA, and the tyrosinase gene begins functioning, turning the mice brown. When the lactose feeding ceases, the mice revert to albinism when they run out of tyrosinase. The modified lac operon might be used to control many other types of vertebrate genes, such as those that are normally lethal early in embryonic development.

An alternative, although even the initial steps of building such construct have yet to be initiated, is to produce a y-aminobutyric acid (GABA) enhancement system to disrupt

GnRH production and also induce sterility. The background and rationale follow.

Recently, GABA has shown great promise as a potential sterilizing agent during embryogenesis. GnRH is the main regulator of gonadal development, and disruption of GnRH production, which can theoretically be accomplished with GABA, could result in enhanced growth, enhanced nutrient utilization, increased carcass yield and improved flesh quality.

GnRH is produced in the neurones of the brain and then secreted into the pituitary gland. This results in the secretion of luteinizing hormone and follicle stimulating hormone into the circulatory system, resulting in steroidogenesis, gametogenesis and growth from the target gonads. Control of GnRH production could be of great utility as it is the key hormone for reproduction, and therefore could also affect other traits that interact with hormone production, sexual maturation, gonadal development and reproduction, such as growth and nutrient utilization.

Production of GnRH is at maximum levels in the pituitary during gamete maturation. GnRH production is stimulated when salmonids become sexually mature, and administration of GnRH can induce oocyte maturation and spawning of fish under artificial conditions.

Perciform fish have three molecular forms of GnRH, chicken GnRH II, salmon GnRH and sea bream GnRH, and these multiple forms have two distinct GnRH neuronal systems, resulting in specific expression in different areas of the brain-forebrain for the fish forms of GnRH and midbrain for the chicken form.

Within the forebrain, the sea bream GnRH is expressed in the preoptic area of the anterior hypothalamus, and the GnRH neurones located here innervate the fish pituitary gland, which is correlated with germ-cell differentiation. Apparently, the salmon GnRH serves another function, as it is expressed in the terminal nerve of the olfactory bulb, and the GnRH nerves here affect sexual behaviour, gonad function and salmon migration, which is reproduc-tively influenced.

The forebrain and midbrain GnRH neurones have different embryonic-stem origins. In mammalian and avian embryos GnRH neurones migrate out of the nasal region into the forebrain, establish their final location and develop projections to the pituitary. Initial studies indicate that the development of these systems in fish is similar to that of mammals and birds. Cells of the preoptic area that express sea bream GnRH in the African cichlid and cells from the terminal nerve of the olfactory bulb that express salmon GnRH both originate from the nasal placode; however, there are temporal differences in expression, with salmon GnRH expression initiated earlier than sea bream GnRH. In mammals and birds, the migration process of GnRH neurones during development takes a few days and is mediated by nerve cells, chemical products and enhancing and inhibiting factors, including GABA, a naturally occurring endogenous compound.

GABA, $C_4H_9NO_2$, molecular weight 103.12, is an amino acid that functions as a neurotransmitter and is a decarboxylation product of glutamate. GABA is a major factor controlling chemotaxic and chemokinetic processes in vertebrate brain development, mediated through GABA-A, GABA-B and GABA-C receptor subtypes on cells of the central nervous system. In mouse, rat and human, GABA-ergic neurones are associated with, are spatially and temporally located near and migrate in a parallel fashion to GnRH neurones. Axonal projections from GABA neurones migrate across the nasal placode and terminate at the cribriform plate, appearing at the nasal side of the cribriform plate at exactly the same time that the GnRH neurones pause at the nasal/forebrain junction. However, at this point the GnRH neurones migrate into the forebrain while the GABA neurones remain behind.

GABA has an inhibitory effect on GnRH neuronal migration. *In vitro* administration of GABA agonists to mouse embryonic nasal explants decreased GnRH gene expression and inhibited normal GnRH neuronal migration to the forebrain by activating GABA-A receptors. GABA directly acts on GABA-A-type receptors to provide a migratory stop signal during mammalian GnRH neuronal development, and the migrational pause is required for proper organization of the GnRH neurones in the forebrain, as *in vivo* administration of GABA-A agonists to early mouse embryos caused decreased migration of GnRH neurones out of the nasal placode, and antagonism of GABA-A receptors caused disorganized distribution of GnRH neurones within the forebrain.

Adult goldfish, rainbow trout and Atlantic croaker respond to *in vivo* administration of GABA with increased plasma gonadotrophin during early gonadal recrudescence; GABA had no effect on gonadotrophin release from dispersed pituitary cells *in vitro*, but pituitary GnRH nerve terminals exhibited an increase in gonadotrophin release, indicating that GABA acts directly on GnRH nerve terminals to potentiate GnRH secretion and subsequent gonadotrophin release. This is further confirmed by the fact that GABA stimulates the release of sea bream GnRH from red sea bream hypothalamic explants. The stimulatory mechanism of GABA on GnRH secretion in fish is mediated by GABA-A-type receptors, just as is observed in mammals.

GABA inhibits activated GnRH neurones in immature mammals and in immature rainbow trout, which do not show an increase in gonadotrophin release unless GABA is coadministered with gonadal steroids. Data to date indicate that interactions of the GABA and GnRH neuronal systems in fish are similar to those in mammals and birds.

In summary, the key factors are that GABA inhibits GnRH neurone migration during embryogenesis in higher vertebrates, and GnRH is critical for gonadal development. This is further substantiated by GnRH mutations in mice and humans, both resulting in hypogonadism and infertility. Introduction of a complete GnRH gene restores fertility in the mutant mouse, and administration of synthetic GnRH restores fertility in the mutant humans. Similarly, recent data indicate that reproductive dysfunction of captive fish can be overcome and early sexual maturation can be induced in fish by the administration of GnRH and GnRH analogues.

The sterilization strategy would require the insertion of two genes. Glutamate is the substrate for the synthesis of GABA. A glutamate construct would need to be transferred to increase levels of glutamate in the developing embryo. Glutamate decarboxylase action would need to be enhanced as well to convert the glutamate to GABA. An artificial glutamate decarboxylase gene would also need to be introduced. The increased expression of both would hopefully result in elevated levels of GABA, which would disrupt GnRH neurone migration, and subsequently the production of GnRH. Fertility of brood fish would be restored by the artificial application of GnRH. The gene action in this case would differ from the sterile feral gene action. Brood stock would need to be homozygous for each transgene. Although both constructs would act as dominant genes, matings of heterozygous individuals would result in some individuals being homozygous recessive for one or both trans-genes and these individuals would be fertile.

Preliminary results where the GnRH anti-sense approach was utilized to transgeni-cally sterilize fish have been very promising. Antisense is most effective when rare messages are targeted and the antisense construct is driven by a strong promoter. A carp P-actin-tilapia salmon-type GnRH antisense construct was injected into Nile tilapia (N. Maclean, personal communication). Transgenic females were crossed with wild-type males. A reduction in fertility of about half that of non-transgenic control females was observed. Fertility was much more greatly reduced in transgenic males crossed to control females. In some cases, 0% fertility was obtained, with an average of about an 80% reduction in fertility. Limited data on transgenic females crossed with transgenic males indicated near-zero fertility.

A tilapia p-actin-tilapia sea bream GnRH antisense construct was injected into Nile tilapia. In this case, no reduction in the fertility of heterozygous transgenic males and females was observed.

Limited data on transgenic females crossed with transgenic males indicated no reduction in fertility. Reciprocal crosses between sea bream and salmon GnRH anti-sense transgenics gave hatch rates that appeared to be dictated by the salmon GnRH antisense parent. Apparently, salmon-type GnRH has a more critical role in fertility than sea bream-type GnRH.

This type of result has been confirmed in transgenic rainbow trout. Transgenic rainbow trout containing salmon-type antisense GnRH from Atlantic salmon, driven by either the GnRH or histone-3 promoter, had reduced levels of GnRH and appeared to be sterile. Preliminary data indicated that spermiation of transgenic males was only obtained after prolonged treatment with salmon pituitary extract, whereas control males spermiated naturally. Data are still needed for the females.

Transgenic sterilization has many applications and would allow domestic, transgenic, hybrid or exotic aquatic organisms to be cultured in any watershed without any potential for affecting native gene pools or local biodiversity. Even conspecific wild strains could be moved from one watershed to another for various purposes without the risk of genetic impact. However, short-term ecological damage could occur from the escape of any of these fish until the time of their death. However, this is the only technology that would assure no genetic impacts. This is the best-case scenario and will provide the best benefit-cost ratio. Another strategy would be to utilize multiple sterile feral constructs for redundant sterility, virtually ensuring no escapement. This technology also allows the protection of proprietary germplasm.

Another transgenic sterilization application is the disruption of reproduction of pest species by the intentional release of sterile individuals into the environment, the 'screwfly approach'. The sterile individuals, males, mate with fertile females, resulting in infertile egg clutches and thus reducing the number of individuals in a population or totally eliminating the population. The release of large numbers of sterile or otherwise unsuitable mates has been used for many years to control insect pests by disrupting reproduction, resulting in infertile eggs. S.A. Davis *et al.* analysed classical deterministic genetic models to determine the result of flooding a target population with individuals carrying a transgene that can be switched on at will by a chemical spray in the watershed. The chemical would induce expression of a trans-gene that causes death or sterility in fish carrying the transgene. The population growth of the target population is reduced, and the triggering chemical would be environmentally harmless. The model indicated that the population level of the target species can be reduced to any desired level, but, if the inducer is applied too often, selection for transgene-free individuals will nullify the effect.

Food Safety of Transgenic Aquatic Organisms

In addition to environmental concerns, the expression of transgenes in foods has brought about concerns and debate regarding food-safety issues. Food safety and education are also critical issues, particularly in regard to consumption of transgenic aquatic organisms. The general public has little understanding of biology and the vagaries of how their food is grown and where it comes from, so public education on the positive aspects of transgenic food and its risks is lacking and is needed.

Food-safety issues posed by transgenic fish are discussed by Berkowitz and Kryspin-Sorensen. Concerns have been voiced over the possible risks of consumption of transgenes, their resulting protein, the potential production of toxins by aquatic transgenic organisms, changes in the nutritional composition of foods, the activation of viral sequences and the allergenicity of transgenic products. These risks have been analysed and, while the majority of genetic modifications to foodstuffs will be safe, the greatest potential for risk and harm is allergenicity.

A transgenic soybean has been developed expressing a gene from Brazil nut to increase its protein content and this transgenic soybean was allergenic to some humans. Labelling laws concerning transgenic foods are also currently being debated and some companies argue that they would be unfairly discriminated against for being compartmentalized as transgenic; however, the potential allergic reactions to transgenic proteins from the donor organisms is one of the strongest arguments for the enactment of some type of labelling.

People do not understand where their food comes from or how it is grown and do not necessarily have a logical perspective, partly because of the influence of the popular media. For instance, people who regularly consume the enzyme chymosin from transgenic bacteria in their Parmesan cheese (the natural source is from calf stomach) or who would use genetically engineered insulin for diabetes are sometimes opposed to the consumption of transgenic aquatic organisms. Hoban and Kendall surveyed North Carolinians in the USA concerning their attitudes towards transgenic foods. Having hypothesized the importance of the perception of biotechnological foods put forward by the media and various watch groups and the importance of the lack of knowledge for influencing opinion, Hoban and Kendall included two biological questions to allow a proper interpretation of the answers. The first was 'Have you ever eaten a hybrid fruit or vegetable?' and the second was 'Do you think it is ethical to eat a hybrid fruit or vegetable?' Approximately 60% of the respondents answered no to both questions. Of course, almost all Americans consume hybrid fruit and vegetables on a regular basis, as most fruits and vegetables are hybrids. There is a great need for public education on biotechnology's advantages, disadvantages, benefits and risks for the benefit of society and to restore faith in science, industry, government and environmental organizations.

International Guidelines

Internationally, the European Union (EU) and the Codex Alimentarius Commission (CAC) have taken the lead roles in voicing concerns about the consumption of transgenic foods and the need for labelling and regulation. CAC is an intergovernmental body established by the Food and Agiculture Organization (FAO) and the World Health Organization, and has a current membership of 163 countries. CAC has developed Codex Standards, which address all food-safety considerations, descriptions of essential food hygiene and quality characteristics, labelling, methods of analysis and sampling and systems for inspection and certification. Codex standards, guidelines and recommendations are not binding on member countries, but are a point of reference for international law.

At the 23rd session in July 1999, the CAC established an Ad Hoc Intergovernmental Task Force on Foods Derived from Biotechnology to develop standards, guidelines or recommendations for foods derived from biotechnology or traits introduced into food by biotechnology. This was to be accomplished on the basis of scientific evidence and risk analysis and having regard to other legitimate factors relevant to the health of consumers and the promotion of fair trade practices.

In the USA, the Food and Drug Administration (FDA) regulates transgenic foods under the purview that the introduction of an exogenous gene is analogous to the introduction of a drug. The consumption of transgenic plants has been approved, but no approval has been granted for the marketing of transgenic aquatic organisms. As stated earlier, an application from Aqua Bounty Farms to market growth hormone (GH)-transgenic salmon in the USA is under consideration by the FDA.

Labelling

The CAC also preliminarily adopted an amendment to the General Standard for the Labelling of Prepackaged Foods. This amendment addresses the need to label foods developed through biotechnology that are substantially different from usual foods. In the EU, the 1997 EU Novel Foods and Novel Food Ingredients Regulation 258/97 dictates mandatory premarket approval and labelling for all foods without a history of consumption in the EU or for food obtained from genetically modified organisms (GMOs). For foods that are substantially equivalent, only a simple notification is required to speed approval if the GMO or product is equivalent to the non-GMO counterpart, and then no extra legislation or oversight is needed. These include cases when there is no evidence of any specific health hazards. The Proposed Draft Recommendations for the Labelling of Foods Obtained through Biotechnology from the CAC states:

> When a food produced by biotechnology is not substantially equivalent to any existing food in the food supply and no conventional comparator exists, the labelling shall indicate clearly the nature of the product, its nutritional composition, its intended use and any other essential characteristic necessary to provide a clear description of the product.

Labelling of GMOs or products from GMOs has become controversial, and major trade conflicts were and are waged between Europe and the USA over labelling of genetically modified crops, such as maize and soybeans. This is extremely important to the USA as 50-70% of its maize and soybean crop is now transgenic. Some countries think labelling is impractical and ambiguous at best; however, most of Europe thinks it is necessary for informed consumer decisions and proper public relations.

Labels can be a potentially positive marketing tool, as in the example of 'dolphin-friendly tuna' and 'organically grown food'. However, the labels could also be used for negative advertising as well. It will be critical to have potentially allergenic transgenic products properly labelled. Additionally, issues such as the extent of the information on the label, verification of the authenticity of labels and enforcement will need to be resolved. The International Federation of Organic Agriculture Movements (IFOAM) has produced standards required for their certification, and vaccines are allowed, but genetically engineered vaccines are not; feeds may not contain GMOs or their products; triploids and genetically engineered species or breeds are also not allowed for organic certification. For many transgenic products, these restrictions do not make biological sense and may be overly restrictive.

References

FAO (2001) Genetically modified organisms, consumers, food safety and the environment. Available at:www.fao.org/DOCREP/003/X9602E/X9602E00.HTM

Kapuscinski, A.R. and Hallerman, E.M. (1990) Transgenic fish and public policy: anticipating environmental impacts of transgenic fish. *Fisheries* 15(4), 2–11.

Klinger, T. (1998) Biosafety assessment of genetically engineered organisms in the environment. *Trends in Ecology and Evolution* 13, 5–6.

Ricker, W.E. (1975) *Computation and Interpretation of Biological Statistics of Fish Population.* Bulletin 191, Department of Environment, Fisheries and Marine Services, Ottawa, Canada.

10

Transgenic Fish: Opportunities and Challenges

As the field of genetic engineering advances, we are beginning to see increased commercial application of this technology. Aquatic animals are being engineered to increase aquaculture production, for medical and industrial research, and for ornamental reasons. While some of these alterations may provide some benefits, the potential effects on human health and the environmental risks that transgenic fish pose to native ecosystems remain unstudied and unknown.

Some of the same genes inserted to provide benefit for transgenic fish may also contribute to higher risk for other species, including humans. Genes inserted to promote disease resistance may allow transgenic fish to absorb higher levels of toxic substances, including heavy metals.In turn, consumers of these fish may be ingesting higher amounts of substances such as mercury and selenium. Transgenic fish that have genes from species such as peanuts or shellfish that are common causes of allergic reactions in humans may prompt allergic reactions in an unsuspecting consumer.

Transgenic species may behave much like invasive species when interacting with the natural environment. They may compete with native species for resources and pose a threat to the genetic diversity of native populations, especially when genetic modifications – such as a rapid growth rate – offer advantages over slower-developing native species. Despite industry assurances that transgenic fish would be unable to naturally reproduce or significantly threaten the environment, some scientists are far more doubtful.

Commercial Applications of Fish Biotechnology

Genetic enhancement programmes and improved germplasms are being applied in both developed and developing countries and are having an impact. Select lines of several aquatic species are utilized in many countries and crossbreeding, particularly of common carp, has been applied in many Asian, European and Middle Eastern countries. Interspecific hybridization, polyploidy, monosex application and transgenic fish are being utilized in various countries worldwide. Breeding companies and aquatic biotechnology companies have been established. Great progress has been made since the early 1970s, but much more can be done.

Polyploidy

Sex-reversal and polyploidy technologies are beginning to have an impact on fish production. This is a distinct advantage for the genetic improvement of aquatic species since terrestrial animals are less plastic genetically and cannot currently be improved via polyploidy. These technologies are on the threshold of advancing global food security significantly, and will be rapidly adopted and applied in both developed and developing countries.

Several examples of the successful application of triploidy exist in aquaculture. Triploid salmon, trout, grass carp and Pacific oysters are commercially produced.

Triploid grass carp are widely utilized in the USA to control aquatic vegetation. Many states have banned the use of fertile diploid grass carp since it is an exotic species, which, if established, might have a detrimental impact on native species and on local aquatic plant ecology. Many states have legalized triploid grass carp because of its sterility and public pressure to utilize these beneficial fish. The triploid genotype results in functional sterility, which allows application of this exotic species for aquatic weed control where diploid grass carp are still illegal. States that allow the use of triploid grass carp require that only triploids be introduced. Screening of most or all potential triploids for ploidy level is required, which makes it desirable for an induction procedure that yields a high percentage of, if not 100%, triploids.

In Europe, particularly the UK, monosex female rainbow trout that are triploids are cultured. The European market requires a larger trout than the American market, and the triploid fish continues to grow and surpass the diploids in size after onset of sexual-maturation effects slows the growth of the diploids. Additionally, as sexual maturation emerges, flesh quality decreases in diploids, whereas triploid salmonids have superior flesh quality at this time. There is also a preference for stocking triploid rather than diploid brown trout in European rivers and lakes for fishery enhancement.

Commercialization of triploid Pacific oysters began on the west coast of the USA in 1985 and they are now widely utilized, accounting for 30% of oyster production. Similar to the example with salmon, the triploid oysters grow faster and have superior flesh quality. The triploidy again counteracts the sexual-maturation effects of decreased growth rate and flesh quality. Triploid oysters are now being used in Brittany, and the demand for triploid Sydney rock oysters is greater than supply. The only oyster with widespread commercial application is the Pacific oyster.

Geographical variation in culturists' attitudes affects the acceptance and utilization of triploid oysters. Triploid Pacific oysters, *Crassostrea gigas,* are commonly cultured on the west coast of the USA and also in France and Australia; however, triploid induction is seldom utilized for eastern oyster, *Crassostrea virginica,* culture on the east coast of the USA, although recently this has started to change.

One of the keys to the widespread utilization of triploid Pacific oysters is hatchery production of seed. There is almost total dependence on hatcheries in the Pacific Northwest for triploid oyster seed. Use of tetraploid males to produce triploid spat is now common. Triploids are especially preferred over diploids in the summer because of the decreased marketability of diploids in spawning condition. Triploid oyster utilization was virtually nonexistent in France until sperm from tetraploids was made available in 1999/2000, resulting in 10-20% utilization of triploid spat. Application of triploid oysters in the UK has not been successful as the growth and meat yield of cytochalasin-B (CB)-pro-duced oysters have been poor; however, hatcheries in the UK appear anxious to try the technology again, using tetraploid sperm. China has produced triploid Pacific oysters since 1997. Worldwide, the application of triploid oysters is minimal except where hatcheries have been established to provide triploid spat.

Sex Reversal and Breeding

Major examples of sex reversal and breeding include the commercial aquaculture of salmonids and Nile tilapia and silver barb farming. Monosex female culture is now being applied in salmonid and silver barb culture, and monosex male culture for Nile tilapia.

Sex-reversed XX male rainbow trout are mated with normal XX female rainbow trout to produce 100% XX monosex females. This technology is commonly used in the UK. The advantages of this system are the faster growth and higher flesh quality of female salmonids. This technology is being adopted by many American rainbow trout farmers. The sex-reversal technology has been adopted, along with triploid technology, in Europe to take advantage of the growth and carcass-quality characteristics of salmonids produced by both of these programmes. Multiple generations of sex reversal and progeny testing of *Oreochromis niloticus* have resulted in both YY males and YY females. When mated, they produce 100% all-male YY progeny. The advantage of this system is the increased growth of the males and the lack of reproduction of the monosex populations. The progeny of the YY males ('genetically male tilapia' (GMT) distinguishes them from sex-reversed male tilapia) have been tested in commercial field trials in the Philippines. Results from on-station trials indicate that GMT application increased yields by up to 58% compared with mixed-sex tilapia of the same strain, which were consistently greater than those for sex-reversed male tilapia. In addition to the negligible recruitment of females in GMT populations, GMT have more uniform harvest size distribution, higher survival and better food-conversion ratios. Economic analysis based on the results of these field trials indicated a major impact on profitability for Philippine tilapia farms from the use of monosex YY populations. Another advantage of monosex technology is the elimination of the need for sex hormones to produce the monosex populations. The reliability of this system is also superior to hybrid technology, where cross-contamination of brood stock of different species living near each other is problematic, resulting in the production of less than 100% male populations. The GMT technique is environmentally friendly, species/strain purity is maintained and the fish produced for culture are normal genetic males.

Although the development process takes several years and is labour-intensive, once developed the production of monosex males can be maintained through occasional feminization of YY genotypes. Assuming that brood-stock purity can be maintained, GMT production can be applied in existing hatchery systems without any special facilities or labour requirements. Disregarding the initial development costs, additional costs for the utilization of GMT technology at the hatchery level should be minimal, while the potential economic advantage to growers is substantial.

The outputs of the research on the YY male technology, GMT and GMT-producing brood stock have been widely disseminated in the Philippines since 1995, in Thailand since 1997 and to a lesser extent in a number of other countries, including Vietnam, China, Fiji and the USA. In the Philippines and Thailand, the strategy for dissemination was to produce and distribute brood stock from breeding centres to accredited hatcheries. As of 2000, 32 accredited hatcheries existed in the Philippines, producing an estimated 40 million GMT, representing 5-10% of fingerling production. In Thailand, conservative estimates indicate that GMT now accounts for more than 15% of fry produced. The XX system was proved effective in maintaining quality control over the technology, but limits the scale of the dissemination. This dissemination strategy is being carried out in a financially viable way, which is considered essential to the long-term sustainability of the technology and its impact.

Dissemination of GMT has reached a scale where the technology may begin to have a significant impact upon worldwide tilapia production. This impact will grow as availability of further improved GMT increases, along with the increased likelihood of restrictions on the use of hormones in aquaculture (combined with consumer resistance).

Wide-scale application of XX all-female (neomale) silver barb has occurred in Thailand. Hatchery trials on station and pilot-scale commercial production in private hatcheries in Thailand demonstrated that neomale brood stock performed satisfactorily. Monosex female fingerlings from neomale brood stock are now produced on a significant scale in commercial hatcheries in Thailand, and research is ongoing in other countries in the region.

Genetic Engineering

Genetic engineering will probably be one of the best tools to improve disease resistance and tolerance of low oxygen in fish in the future. Faster-growing, genetically engineered fish will soon be available for commercial use. However, environmental concerns, public fear, and government regulation may slow the commercialization of genetically engineered fish. Transgenic fish must be shown to be safe for the environment and for consumption. If this is demonstrated, the general public, leaders and legislators must be educated on the safety and benefits of genetically engineered food, and scientists, industry and policy makers need to work together to ensure realistic and reasonable government regulations.

Recently, salmon have been produced in Europe, New Zealand and North America that contain additional salmon growth hormone (GH) genes, which enhance levels of GH and rate of growth. However, approval and consumption of transgenic salmon will probably be long in coming in Europe because of the strong anti-transgenic sentiment there. Otter Ferry Salmon in Scotland initiated a growth trial with transgenic Atlantic salmon in 1996 in a closed system. The fish were grown for 18 months and then destroyed, and the Scottish Salmon Association distanced itself from the experiments, fearing consumer protests in the market. Later, when it was learned in the House of Commons that the government had approved the privately funded experiment with transgenic salmon, there was great concern and opposition to this government support in the UK. The trials were successful as the transgenic salmon grew at four times the rate of controls; however, nine salmon-growing countries agreed to a ban on genetically modified fish subsequent to this trial.

Possibly, transgenic carps have been commercialized in China and transgenic Nile tilapia in Cuba, but this is disputed. The University of Connecticut is working with Connecticut Aquaculture to commercially produce transgenic tilapia (*Oreochromis* sp.), and Kent SeaFarms (San Diego, California) has started a programme for developing transgenic striped bass, although products from these efforts have yet to be commercialized.

Issues in Maintaining Genetic Quality

An important aspect of genetic biotechnology is not only improvement of performance, but also the maintenance of genetic and stock quality. Inbreeding is the mating of individuals more closely related to one another than the average of the population. Inbreeding increases homozygosity, which can lead to reduced rate of growth, viability and reproductive performance and increased biochemical disorders and deformities from lethal and sublethal recessive alleles. Inbreeding can occur in both cultured and natural populations. When compared with much larger populations in Ukraine, the Hungarian meadow viper, *Vipera ursinii rakosiensis,* had low major histocompatibility complex (MHC) variability, which presumably reduced the immune response, and greater genetic uniformity. These small populations also had birth deformities, chromosomal abnormalities and low juvenile survival, suggesting that the Hungarian vipers are exhibiting inbreeding depression.

Avoidance of inbreeding and utilization of breeding schemes to avoid inbreeding are critical for the maintenance of genetic variance in both developed and developing countries; it may be just as important not to lose production from lack of genetic maintenance as to gain production from genetic enhancement. This is especially problematic when farming species with high fecundity. such as Indian and Chinese carps, where only a few brood stock can easily meet fry-production needs. The detrimental effects of inbreeding depression are expressed when the inbreeding coefficient, F, reaches approximately 0.25.

Regardless of whether or not a commercial company or farm develops or obtains its own genetically enhanced aquatic organism or does not have an improved stock, it is important to maintain stock quality and prevent the current stock from regressing in genetic quality. Even when genetic improvement is not the goal, the avoidance of inbreeding depression and random genetic drift primarily revolves around population size.

The objective is to prevent the inbreeding coefficient, F, of a population from reaching 0.25 or 25%. At this level of inbreeding, inbreeding depression is likely to occur in fish.

A fish farmer can easily calculate the accumulation of inbreeding per generation as follows:

> ΔF/generation = 1/8 (number of males that successfully breed) + 1/8 (number of females that successfully breed)

This indicates the change in F each individual generation and must be calculated for each generation. The values are then added from all generations to determine the accumulated inbreeding to the present. Generally, if 50 pairs of fish randomly mate each generation, an F of 0.25 will not accumulate for 50 generations. Most large channel catfish farms, *Clarias* farms and tilapia farms maintain larger brood populations than 50 pairs, and inbreeding depression is probably not a concern in these cases. Of course, even with an adequate population size, inbreeding could still occur if progeny from these 50 pairs are not mixed or become segregated and care is not taken to randomly choose replacement brood stock that would represent all spawns from the 50 mat-ings. Also, selection programmes are actually mild inbreeding programmes, as mating is not random and, even with large population sizes, inbreeding could, theoretically, occur. Of course, when inbreeding is known, suspected or calculated, it and its effects can be immediately eliminated by mating to unrelated individuals and F immediately returns to zero.

Avoidance of inbreeding and maintenance of genetic quality can be especially critical for genetic approaches to biotechnology. The nature of these biotechnologies makes it tempting to initiate populations with inadequate or small population numbers. This could lead to inbreeding or drift, which might partially or completely negate or counter the improvements made through the biotechnology. Gene transfer, gynogene-sis, cloning, sex reversal and breeding would all be potentially initiated with small population numbers without extra effort. Of course, these genetic changes could be bred into a population and genetic variation and population-size parameters addressed through traditional means. However, this could delay commercialization by a generation while the new genotype is being spread into a larger population base. However, another strategy would be to initiate commercialization with the narrower genetic base and correct the problem over time with conventional breeding strategies.

Bilateral asymmetry—unbalanced numbers for meristic traits on the right and left halves of the body—has been shown to be an indirect indicator of homozygosity in fish. Since the two halves of an embryo should be genetically identical, the random asymmetry can be interpreted as failure of the genotype to adequately regulate the development of the phenotype. It is assumed that this bilateral asymmetry may also be an indicator of inbreeding depression and reduced fitness. Bilateral asymmetry of abdominal bristles in 32 fruit-fly lines was not correlated to two measures of fitness: productivity (a combined measure of fecundity and egg-to-adult survivorship) and competitive male mating success. This indicates that, although bilateral asymmetry is usually not desirable, it may not always be a reliable indicator of reduced fitness.

Bryden and Heath found that bilateral asymmetry for the meristic and quantitative traits in chinook salmon is not heritable, and they conclude that bilateral asymmetry estimates in chinook salmon will not be confounded by appreciable additive genetic contributions and thus can be reliably used as an environmental and genetic stress indicator. This should also indicate that bilateral asymmetry has a dominance genetic component.

Allozymes do not always measure changes in inbreeding accurately and this was the case in guppy *Poecilia reticulata*. Three closed lines of guppies (n = 10) were propagated for six generations, achieving inbreeding levels of 0.19, 0.32 and 0.41 and salinity tolerances were 82.5, 71.7 and 67.6%, respectively, of their original values. Salinity tolerance was reduced linearly by 8.4% per 10% increase in inbreeding.

To monitor parentage to maintain genetic quality, the number of markers necessary to accomplish the goal needs to be known. The analysis of Bernatchez and Duchesne indicates that, if sufficient allelic diversity exists, a relatively low number of loci is required to achieve high allocation success to full-sib families even for relatively large numbers of possible parents, but there is no significant gain in increasing allelic diversity beyond approximately six to ten alleles per locus in population-assignment studies.

Microsatellites are an excellent tool for checking pedigrees and lineages. Cunningham *et al.* examined Thoroughbred horse pedigrees extending back about 300 years with 12 microsatellite loci, and agreement between relatedness calculated from pedigree records and estimated from the microsatellites was extremely high.

Theoretically it might be a good idea to deliberately inbreed small, captive populations to eliminate inbreeding depression as inbreeding exposes deleterious recessive alleles in the homozygous state to selection, allowing their removal and rapid purging from the population. This would eliminate future inbreeding depression even in small populations. Frankham *et al.* developed two populations of fruit flies by inbreeding (brother-sister mating) for 12 generations. One population was outbred until the inbreeding started and was presumably genetically variable, while the second type had been inbred for 20 generations and then multiply hybridized just prior to the final inbreeding experiment. Thus the second population was also genetically variable, because of the hybridization, but purged of deleterious alleles because of the prior inbreeding. The purging was not successful, and there was a small but non-significant difference between the extinction rates at an inbreeding coefficient of 0.93 in the non-purged (0.74 ± 0.03) and purged (0.69 ± 0.03) treatments. This is consistent with other evidence indicating that the effects of purging are often small, and purging using rapid inbreeding in very small populations is not a reliable method to eliminate the deleterious effects of inbreeding.

Results of purging in the laboratory may differ from those in more natural conditions. The native silversword, *Argyroxiphium sand-wicense sandwicense,* plants on the Hawaiian island of Mauna Kea were almost eliminated by cows and then were reintroduced in a recovery programme. The population crash did not reduce the number of microsatellite alleles or heterozygosity. However, a population bottleneck resulted from the reintroduction, reducing the observed number of alleles, effective number of alleles and expected heterozygosity, though not the observed heterozygosity. The data are consistent with other studies indicating that small populations, including those resulting from severe reductions, are not necessarily devoid of genetic variability, and distortion can be caused when a population is founded or supplemented by a few individuals.

Genotype-environment interactions may occur for inbreeding depression, and laboratory studies, where organisms are pampered, may not reflect the inbreeding depression observed in more natural conditions. Inbred *Mus domesticus* males sired only one-fifth as many surviving offspring as outbred males because of their poor competitive ability and survivorship in large, seminatural enclosures, but this reduced fitness is not observed in the laboratory where competition for mates does not exist, or in inbred females, which do not have to compete for mates in either environment. The reduced fitness of inbred males in the competitive environment was almost five times larger than what has previously been seen in the laboratory.

Theoretically, the mixing of sperm from multiple males, followed by fertilization of eggs, should increase genetic diversity. However, in reality, this is not the case. When salmon milt was mixed, sperm from one male were more competitive and dominated the fertilization, so fertilization was essentially the same as using a single male. Ten microsatellite loci were analysed for the veliger larva of abalone; three loci were generally sufficient to assign parentage. Most fertilizations were attributable to just one of the males in a multiple male sperm pool. Fertilizing with a mixed pool of sperm is not a good procedure for increasing genetic diversity, because almost always one male has faster, more competitive sperm, which fertilize the majority of, if not all, the eggs. The practical solution is to divide the eggs from single females into multiple lots, and then fertilize each lot with sperm from different single males to eliminate competition of the sperm and to maximize genetic diversity.

Similar results were obtained with Pacific oysters. Boudry *et al.* used a single, highly variable, microsatellite locus to trace the sources, showing large variation in parental contributions at various developmental stages of Pacific oysters, which led to a strong reduction of effective population sizes in experimental populations. Segregation distortions fluctuated with time, and variance in reproductive success was attributed to gamete quality, sperm-egg interaction and differential viability among genotypes. Sperm competition increased reproductive variance and decreased effective population size, as observed in abalone and salmon.

Waldbieser and Wolters studied reproductive behaviour in channel catfish using microsatellite markers. Multiple spawning by males was found in seven of the eight ponds examined, and 47% of the males fertilized two to six egg masses over 1-8 weeks, although most were 2-3 weeks apart. Four different spawns were identified as two full-sib families and were probably due to interrupted spawning, with the pair reinitiating mating at a later time and/or different location. The fish did not randomly spawn. A subset of males would dominate spawning for a couple of weeks and then they would be replaced by a new dominant subset.

Microsatellite analysis indicates that Atlantic salmon are polygynous, and polygamous mating behaviour is the norm. The number of offspring, the major component of fitness, was correlated with the number of mates, but not with body size. This seems almost contradictory to hatchery results, where artificial mixing of sperm does not result in a strong genetic contribution of multiple males. Perhaps, under the natural conditions, the stimulation of natural mating results in more even maturation of sperm among the males.

Some controversy exists concerning the effectiveness and benefits of supplemental stocking programmes. Hedrick *et al.* evaluated the effectiveness of a supplementation programme for winter-run chinook salmon in the Sacramento River in California. In this well-designed programme, supplementation stabilized and increased the effective population size. Returning spawners represented a broad sample of parents and not fish from only a few families.

Ryman and Laikre demonstrated that the effective population number of an endangered, wild population would actually be decreased by supplementing it with hatchery stock based on the first generation of stock supplementation; however, Wang and Ryman show, both analytically and by simulation, that the reduction in effective number (both drift and inbreeding) is quickly reversed in later generations as long as the census size of the population -the actual number of animals, including those breeding in the hatchery and the wild—is increased by supplementation. Selecting either wild or hatchery brood stock accomplished the increase in effective population size. This experiment also illustrates the point that erroneous conclusions can be reached in short-term experiments of insufficient length.

Poon and Otto developed a model to determine the balance of detrimental and beneficial mutation, and the modelling exercise indicated that the loss of fitness caused by random fixation of deleterious mutations is directly proportional to the dimensionality and inversely proportional to the effective population size. Poon and Otto conclude that the reciprocal relationship between the loss of fitness and effective population size implies that the fixation of deleterious mutations is unlikely to cause extinction when there is a broad scope for compensatory mutations, except in very small populations, and pleiotropy plays a large role in determining the extinction risk of small populations. This appears logical as, when one gene affects more than one trait, its importance and impact must be substantial for fitness. Compensatory mutations must be considered to increase the accuracy of predictions concerning the genetics of small populations.

To minimize the accumulation of inbreeding, each brood fish should contribute equally to the next generation, but this is often impossible to accomplish. Sonesson and Meuwissen developed a scheme of overlapping generations to minimize inbreeding and evaluated it with simulation. Selection of older breeders from the age-class mixture reduced the rate of accumulation of co-ancestry and inbreeding, was practical when mature mortality rates are reasonably low and there is good control over family size and, theoretically, was ever superior to pedigree analysis for reducing the accumulation of inbreeding. However, surely a combination of this overlapping-generation technique and pedigree maintenance would be even more effective.

Standard management practices recommend equalizing family sizes from one-on-one mating of males and females to minimize inbreeding and random genetic drift, but theoretically this procedure can also reduce the intensity of natural selection (actually domestication) for fitness as mutational load will accumulate. However, computer simulations indicate that equalization of family sizes does not produce a high short-term threat to small conserved populations, up to about 20 generations, and the more efficient preservation of genetic variability outweighs the disadvantages of the procedure.

Doyle *et al.* describe a method to increase the genetic diversity of a bottle-necked brood stock without bringing in new breeders. The procedure is an extension of minimal kinship selection, which has been used to preserve diversity in brood stocks where complete pedigree records exist, such as in zoos. Unfortunately, pedigrees are rarely available in fish hatcheries; therefore, instead of using pedigrees to calculate kinships, Doyle *et al.* used Ritland's genetic related-ness estimator, calculated from microsatel-lites, to estimate the mean relatedness of each potential breeder to the whole population. A subset of breeders was then selected to maximize the number of founder lineages, in order to carry the fewest redundant copies of ancestral genes. Microsatellite data from a hatchery population of red sea bream for which pedigrees were independently available were used to validate the method, resulting in higher standard measures of marker diversity in the selected subset of breeders than in randomly chosen subsets. In the next generation, a partial reversal of the effects occurred from genetic erosion and drift. The procedure differs from marker-assisted selection (MAS) in that DNA marker data were used to identify rare pedigrees or extended families, rather than to identify rare chromosome segments carrying quantiative trait loci (QTLs), and the particular application emphasized the recovery of the genetic diversity lost when a hatchery is founded with a small and non-representative sample of an ancestral wild population.

Inbreeding and thus selection could change genetic covariances. Phillips *et al.* produced a large number of small populations of *Drosophila* to generate random drift, resulting in no change in average variance-covariance matrix (G) and overall structure among traits of the inbred lines relative to the outbred controls. However, there was a great deal of variation among inbred lines around this expectation, even changes in the sign of genetic correlations. Since any given line can be quite different from the outbred control, it is likely that in nature unreplicated drift will lead to changes in the G matrix; the shape of G is malleable under genetic drift, and the evolutionary response of any particular population will probably depend on the specifics of its evolutionary history. Thus, the establishment of a small aquaculture population from a larger population may result in genetic changes that are qualitatively and quantitatively different from those in the source brood stock.

Amos *et al.* examined neutral markers in three long-lived vertebrates -long-finned pilot whale, grey seal and wandering albatross—revealing negative relationships between parental similarity and genetic estimates of reproductive success. The negative correlation between parental relatedness and fitness was not merely the result of inbreeding depression, caused by the mating of close relatives, as the correlation extended to low levels of relatedness, where conventional inbreeding depression was unlikely. There appears to be a positive advantage to finding a dissimilar mate, and this result was similar to that for Atlantic salmon, which may choose mates to maximize MHC diversity.

Genetic Conservation

Gene banking or genetic conservation may be accomplished by protecting native populations, by artificially propagating a variety of genotypes, lines and strains and/or through the use of cryopreservation. Cryopreservation of sperm from several fish species has been accomplished and is fairly routine; however, the technology for the cryopreservation of eggs and embryos of aquatic species does not yet exist. An alternative would be to cryopreserve cells from the blastula stage of the embryo. After thawing, these cells can be placed into a donor embryo of the same species. If the donor is triploidized, the only viable cells in the gonads will be from the cryopreserved blas-tulas and, reproductively the entire genome of an individual has been preserved and banked, including the mitochondria and the cytoplasm. Of course in the case of cryopreserved sperm, even if the individual is regenerated via androgenesis, the cytoplasm and the mitochondrial DNA (mtDNA) from the original fish are lost, so the exact total genome and its potential interactions with the cytoplasm have been lost.

Currently, long-term storage of fish eggs or embryos is not possible, severely limiting the effectiveness of cryopreservation for gene banking. However, androgenesis is a potential mechanism to recover diploid genotypes from cryopreserved sperm, which is, of course, a viable technology. There are major drawbacks to this strategy since androgens have low viability, all individuals would be 100% homozygous and the mtDNA portion of the genome would be different from the original cryopreserved individuals unless donor eggs came from exactly the same family or genotype.

Cloning of individuals from somatic cells is another option for regenerating individuals. The problem is that, in the case of cloning, although it would theoretically be superior to the androgenesis approach because androgens are 100% homozygous and clones would preserve the genetic variation from every locus of the individual, clones, like the androgens, have the cytoplasm and mtDNA of the recipient cell and not their own. Cloning has become a reality.

Another potential method of conserving a species may be the isolation of spermatogonial stem cells, if they can be adequately frozen, and transferring these cells into other species, which have been made experimentally, sterilized and depleted of cells in their testis. The spermatogonial stem cells then colonize in the host male and differentiate into sperm cells, which may then be used for fertilization. Stem cells have been cryopreserved and revived and they behave similarly to those of non-cryopreserved controls. Again, the shortcoming is the cytoplasm and mtDNA issue. Similarly, ovarian stem cells could be transplanted to regenerate ovarian tissues, and theoretically this would not have the disadvantage of loss of original mtDNA genotype and cytoplasm.

Another potential technique to gene-bank individuals or transfer genes might be the transfer of genes into cultured blastula cells and then reinserting transgenic or normal blastula cells into embryos via nuclear transplantation. Alternatively, diploid blastulas, transgenic or normal, could be inserted into triploids and, if they became part of the germ line, a fertile individual would be regenerated. Chen *et al.* were able to transplant 59th-generation cultured blastula cells into crucian carp. Nuclear-transplant gastrulae were obtained at a rate of 7.9%. Cells from these gastrulae were then serially transplanted into additional enucleated crucian carp cells and one adult fish was obtained that lived for 3 years. Unfortunately, this fish was aneuploid, illustrating some of the problems of nuclear transplantation with cultured cells. The gonads of this fish did not develop.

The same researchers also transferred cultured kidney cells in a similar manner and produced a fertile female; however, they do not provide sufficient information to confirm that the injected nuclei directed development. Gasaryan *et al.* irradiated and then transferred blastulae of loach, *Misgurnus fossilis,* donors, and the resulting nuclei were 1N, 2N, 3N, 4N. Nuclear transplantation is somewhat problematic.

Tufto addresses the possibility of re-establishing an extinct population using animals bred in captivity, considering that those animals may have strayed from the local fitness optimum. The supplementation of wild stocks by release of domesticated, gene-banked animals could damage the fitness of the remnant wild population. The deterministic model he developed had complex results, dependent on selection intensity, immigration rate, recombination rate, the presence or absence of competitors, the basal intrinsic rate of increase and the carrying capacity. When immigration and selection were both small, the population was reduced below carrying capacity only if the immigrating maladapted animals were far from the optimum, indicating that it is important to prevent as much domestication in the gene bank as possible. When selection is strong but density dependence is weak, as it might be in the early stages of reintroduction, it may be feasible to use maladapted individuals to initiate a population. If the ongoing supplementation (immigration) continues at a low, constant rate, the new population will adapt sufficiently and quickly reach a stable population equilibrium.

Constraints and Limitations of Fish Biotechnology

During the International Conference on Aquaculture in the Third Millennium, a genetics working group convened to define the issues, constraints and limitations facing aquaculture genetics in the near future. Working-group members included

Rex A. Dunham (chair), Kshitish Majumdar (co-chair), Zhanjiang (John) Liu, Eric Hallerman, Gideon Hulata, Trgve Gjedrem, Peter Rothlisberg, David Penman, Nuanmanee Pongthana, Ambekar Eknath, Modadugu Gupta, Graham Mair, P.V.G.K. Reddy, Janos Bakos, S. Ayyappan, Devin Bartley and Gabriele Hoerstgen-Schwark. The following is adapted from Dunham *et al.* and this was the output from this working group regarding these issues.

Several constraints and limitations will need to be overcome for aquaculture genetics to have its maximum impact and benefit in the coming years. These include environmental issues, such as biodiversity, genetic conservation and the environmental risk of genetically altered aquatic organisms; research issues, such as funding, training of scientists and impact assessment; economic issues, such as proprietary rights, dissemination, food safety and consumer perceptions; and political issues, such as government regulation and global cooperation.

Research Issues

Manpower—the lack of trained traditional and molecular aquaculture geneticists—is another potential constraint that needs to be addressed. Impact assessment is another void that needs more intense consideration. This is needed to ensure that the research and germplasm developed through research are appropriate for the commercial sector and are applied properly and disseminated properly to achieve maximum impact. Research on impact assessment should be influential in the design of the most effective breeding programmes, extension development and dissemination strategy.

General Recommendations

The genetic improvement of cultured fish and shellfish should be given higher priority by government, non-government and commercial organizations. This will increase productivity and turnover rate, result in better utilization of resources and reduce production costs. Multiple-trait selection programmes need to be further developed. Efficient breeding plans should be developed where selection is combined with other genetic technologies. Better genetic controls need to be developed for monitoring the progress of breeding programmes. More education and training programmes are needed for the further development of aquaculture geneticists, especially in developing countries. The establishment of national and international genetic controls, including homozygous and heterozygous clonal populations for some key species, would help in comparing genetic results and germplasm from different research institutions, increase global cooperation and enhance research efficiency. Domestication of some wild-cultured organisms, such as shrimp, is needed.

Development Issues

There is a greater need for intervention and collaboration in developing hatchery management and breeding programmes for low value/low input species in developing countries. Good brood-stock management needs to be promoted to counter the negative genetic impacts of inbreeding. Species and traits relevant to low-input systems need to be prioritized for genetic-enhancement programmes that better address food-security issues.

Networking as a way to circumvent the lack of resources (human and infrastructure) in developing countries should be strengthened to coordinate sharing of information, expert opinion, education and research and to assist in obtaining funding. Efforts should be made to design and promote equitable dissemination strategies that ensure that genetic enhancements have positive impacts on aquaculture and food security and enhance livelihoods. Research should be carried out to assess the impact of research, development and dissemination of genetically improved stocks, including intellectual property rights (IPR) issues. Species and traits need to be prioritized for genetic-enhancement programmes that better address global food security.

Biodiversity Issues

Aquatic biodiversity needs to be characterized and protected.

The population genetics of many key species require closer examination. Interactions of wild and domesticated species needs much closer study, including modelling. There should be an intensification of live, frozen and molecular gene-banking efforts. More research is needed in the area of effective sterilization techniques for domesticated and transgenic aquatic organisms. There is a need for greater controls of trans-boundary movements of aquatic germplasm.

Research on transgenic aquatic organisms should continue because of their potential benefits (especially in developing countries); however, much greater understanding of potential environmental impacts is necessary. Linkages should be formed among civil society, organizations, scientists, industry and governments to address genetic issues and to support the development of practical regulations and sound policy. Dissemination of transgenic aquatic organisms for aquaculture should only be carried out within the framework of adequate regulations and policy.

Political Issues

Worldwide, policies for research and marketing of transgenic food organisms range from non-existent to stringent, as in the EU. Government regulation of transgenic aqua-cultured species, based on sound scientific data, is lacking and much needed. Not surprisingly, global cooperation on issues of biotechnology is not unified. Countries party to the International Convention on Biological Diversity (CBD) and involved in the World Trade Organization (WTO) are divided on key issues, such as transport of transgenic organisms between countries, precautionary principles driving biosafety decisions, liability in the case of negative effects on human health or biodiversity, possible social and economic impacts on rural cultures, regulation of transgenic products across borders, food safety and protection of transgenic trade goods. Recently, however, international legislation, guidelines and codes of conduct have been, or are being, established to help address these areas of concern.

Economic Issues

Among the key issues are economic ones, particularly those revolving around proprietary rights issues. There are many aspects to this, including those related to biodiversity and molecular genetics, as organisms found in almost any country have genes that are potentially valuable to another organism in a different country. Ownership in cases of international germplasm transfer is an issue. A genetics research and breeding programme requires financial support. Appropriate, equitable dissemination and ownership of germplasm developed with tax monies or donor funding, with the goal of having a positive impact on the impoverished in developing countries, is a complex and controversial topic. This is an increasingly difficult problem as, with the initiation of private breeding companies and biotechnology companies, alternative options to government dissemination, with both impact and income generation for research opportunities, exist. The most cost-efficient dissemination strategies with the highest impact have not been completely defined and evaluated.

References

Chen, T.T. and Powers, D.A. (1990) Transgenic fish. *Trends in Biotechnology* 8, 209–214.

Glick, B.R. and Pasternak, J.J. (1998) *Molecular Biotechnology: Principles and Applications of Recombinant DNA*. ASM Press, Washington, DC.

Hoban, T.J. and Kendall, P.A. (1993) *Consumer Attitudes about Food Biotechnology*. Department of Sociology and Anthropology, North Carolina State University, Raleigh, North Carolina

Maclean, N., Penman, D. and Talwar, S. (1987a) Introduction of novel genes into fish. *Biotechnology* 5, 257–261

Murray, D.R. (1991) *Advanced Methods in Plant Breeding and Biotechnology*. CAB International, Wallingford, UK.

Bibliography

Avise, J.C. (1994) *Molecular Markers, Natural History and Evolution.* Chapman & Hall, London.

Battacharya S., Dasgupta, S, Datta, M and Basu, D., (2002). Biotechnology input in fish breeding. *Indian journal of biotechnology,* 1:29-38.

Beaumont, A.R. & Fairbrother, J.E. (1991) Ploidy manipulation in molluscan shellfish: a review. *Journal of Shellfish Research,* 10, 1–18.

Beaumont, A.R. (ed.) (1994) *Genetics and Evolution of Aquatic Organisms.* Chapman & Hall, London.

Benzie, J.A.H. (ed.) (2002) *Genetics in Aquaculture VII.* Elsevier, Amsterdam.

Bidwell, C.A., Chrisman, C.L. and Libey, G.S. (1985) Polyploidy induced by heat-shock in channel catfish. *Aquaculture* 51, 25–32.

Brown, T.A. (1999) *Genomes.* Bios Scientific Publishers, Oxford.

Carvalho, G.R. & Pitcher, T.J. (eds) (1995) *Molecular Genetics in Fisheries.* Chapman & Hall, London.

Chen, T.T. and Powers, D.A. (1990) Transgenic fish. *Trends in Biotechnology* 8, 209–214.

Chen, T.T., Zhu, Z., Lin, C.M., Gonzalez-Villasenor, L.I., Dunham, R. and Powers, D.A. (1989) Fish genetic engineering: a novel approach in aquaculture. In: *Aquaculture '89 Proceedings.* National Shellfish Association, Los Angeles, California

Dunham, R.A. (1990b) Genetic engineering in aquaculture. *AgBiotech News and Information* 2, 401–406.

Dunham, R.A. and Liu, Z. (2002) Gene mapping, isolation and genetic improvement in catfish. In: Shimizu, N., Aoki, T., Hirono, I. and Takashima, F. (eds) *Aquatic Genomics: Steps Toward a Great Future.* Springer-Verlag, New York, pp. 45–60.

Falconer, D.S. (1989) *Introduction to Quantitative Genetics,* 3rd edn. Longman, New York.

FAO (2001) Genetically modified organisms, consumers, food safety and the environment. Available at:www.fao.org/DOCREP/003/X9602E/X9602E00.HTM

Fletcher, G. and Davies, P.L. (1991) Transgenic fish for aquaculture. *Genetic Engineering* 13, 331–369

Gjedrem, T. (2000) Genetic improvement of cold-water fish species. *Aquaculture Research,* 31, 25–33.

Glick, B.R. and Pasternak, J.J. (1998) *Molecular Biotechnology: Principles and Applications of Recombinant DNA.* ASM Press, Washington, DC.

Hillis, D.M. & Moritz, C. (eds) (1996) *Molecular systematics,* 2nd edn. Sinauer Associates Inc., Sunderland

Hoban, T.J. and Kendall, P.A. (1993) *Consumer Attitudes about Food Biotechnology.* Department of Sociology and Anthropology, North Carolina State University, Raleigh, North Carolina

Kapuscinski, A.R. and Hallerman, E.M. (1990) Transgenic fish and public policy: anticipating environmental impacts of transgenic fish. *Fisheries* 15(4), 2–11.

Karunasagar, I. (1999). Diagnosis treatment and prevention of microbial diseases of fish and shellfish. *Curr.Sci.,* 76:387-399.

Kearsey, M.J. & Pooni, H.S. (1996) *The Genetical Analysis of Quantitative Traits.* Chapman & Hall, London.

Klinger, T. (1998) Biosafety assessment of genetically engineered organisms in the environment. *Trends in Ecology and Evolution* 13, 5–6.

Lakra, W.S. and Das ,P., (1998). Genetic engineering in aquaculture, *Indian .J.Anim.Sci.,* 68(8);873-879.

Liu, Z.J. and Dunham, R. (1998a) *Genetic Linkage and QTL Mapping of Ictalurid Catfish.* Circular Bulletin 321, Alabama Agricultural Experiment Station, 19 pp.

Lutz, G.C. (2002) *Practical Genetics for Aquaculture.* Blackwell Science, Oxford.

Maclean, N., Penman, D. and Talwar, S. (1987a) Introduction of novel genes into fish. *Biotechnology* 5, 257–261

Majerus, M., Amos, W. & Hurst, G. (1996) *Evolution: the Four Billion Year War.* Longman, New York.

Meyer, A. (1993). Evolution of mitochondrial DNA in fishes. In: *Biochemistry and Molecular Biology of Fishes,* V. 2, Hochachka, P.W. and Mommsen, T.P. (eds). Elsevier, Amsterdam, pp 1-38

Morizot, D.C. and Siciliano, J.J. (1984) Gene mapping in fishes and other vertebrates. In: *Evolutionary Genetics of Fishes*. Monograms in Evolutionary Biology, pp. 173–233.

Murray, D.R. (1991) *Advanced Methods in Plant Breeding and Biotechnology*. CAB International, Wallingford, UK.

Nakamura, Y., Leppert, M., O'Connell, P., Wolff, R., Holm, T., Culver, M. and Martoin, C. (1987). Variable numbers of tandem repeats (VNTR) markers for human gene mapping. *Science*, 235: 1616-1622

Purdom, C.E. (1973) Induced polyploidy in plaice (*Pleuronectes platessa*) and its hybrid with the flounder(*Platichythys flesus*). *Heredity* 29, 11–24.

Ricker, W.E. (1975) *Computation and Interpretation of Biological Statistics of Fish Population*. Bulletin 191, Department of Environment, Fisheries and Marine Services, Ottawa, Canada.

Ryman, N. & Utter, F. (eds) (1987) *Population Genetics and Fishery Management*. University of Washington Press, Washington.

Schultz, R.J. (1980) Role of polyploidy in the evolution if fishes. In: Lewis, W.H. (ed.) *Polyploidy: Biological Relevance*. Plenum Press, New York.

Tave, D. (1993) *Genetics for Fish Hatchery Managers*, 2nd edn. Van Nostrand Reinhold, New York.

Tiews, K. (ed.) *Selection, Hybridization and Genetic Engineering in Aquaculture*, Vol. 2. Heeneman Verlagsgesellschaft, Berlin, pp. 239–253

Turner, P.C., McClennan, A.G., Bates, A.D. & White, M.R.H. (1997) *Instant Notes in Molecular Biology*. Bios Scientific Publishers, Oxford.

Utter, F.M. (1991). Biochemical genetics and fishery management: an historical perspective. *Journal of Fish Biology*, supplement A, 39: 1-20.

Ward, R.D., Skibinski, D.O.F. & Woodwark, M. (1992) Protein heterozygosity, protein structure and taxonomic differentiation. *Evolutionary Biology*, 26, 73–159.

Whitehead, P.K. (1994) Mapping and electroporation of mitochondrial DNA from three species of ictalurid catfish. Doctoral dissertation, Auburn University, Auburn, Alabama, USA.

Winter, P.C., Hickey, G.I. & Fletcher, H.L. (1998) *Instant Notes in Genetics*. Bios Scientific Publishers, Oxford.

Wright, J.M. (1993). *DNA fingerprinting of fishes. In: Biochemistry and molecular biology of fishes*, Hochachka and Mommsen (eds). Elsevier Science Publishers, B.V., pp. 57-91

Index